MASSACRE

The Goliad Witnesses

Edited and Compiled by Michelle M. Haas

Copano Bay Press
2014

ISBN 978-1-941324-02-8

Contents

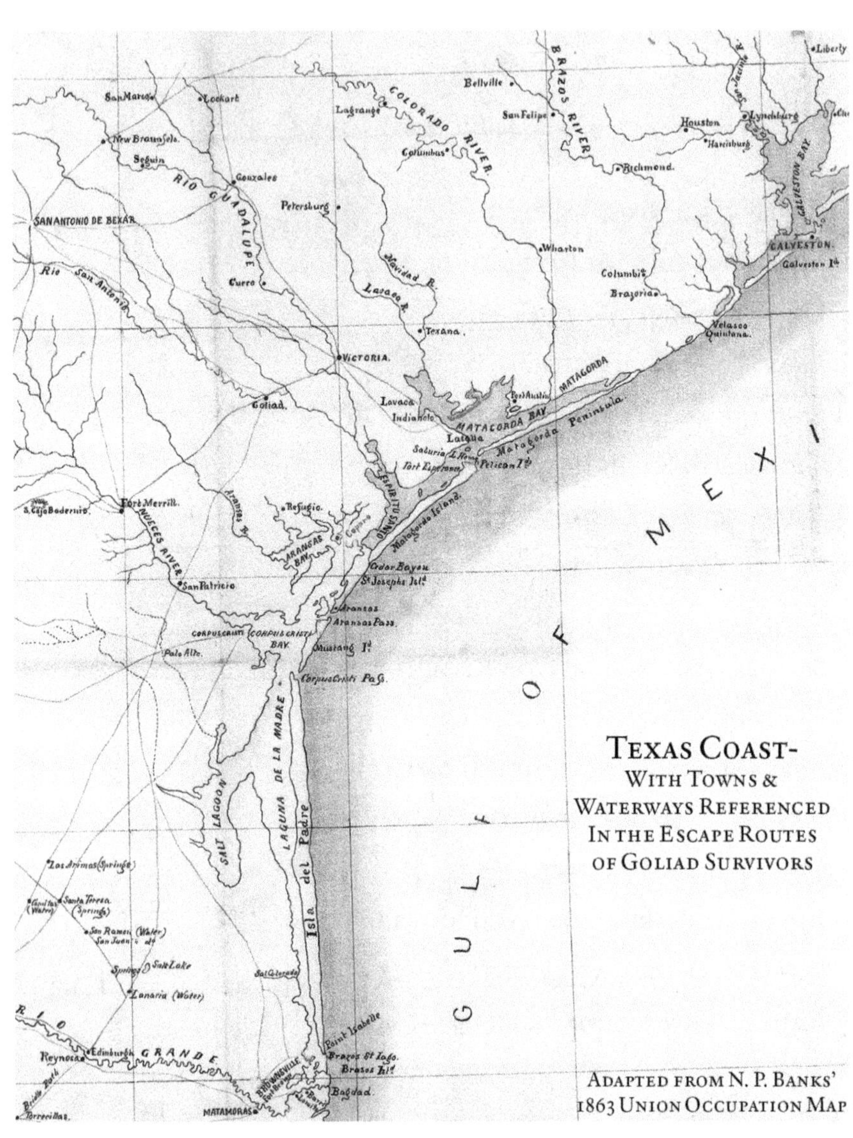

Texas Coast-
With Towns & Waterways Referenced In the Escape Routes of Goliad Survivors

Adapted from N. P. Banks' 1863 Union Occupation Map

Publisher's Note

On Palm Sunday, March 27, 1836, 342 men were slaughtered in and around Presidio La Bahia, at the order of Santa Anna. Most were men who surrendered under their commander, Colonel James Fannin, a week prior following the Battle of Coleto. Twenty-eight escaped the massacre and another seventeen were spared to be used as physicians, orderlies and tradesmen. This volume is intended to give voice to as many of the survivors of the massacre as are extant. Of those who were spared, four are here. Of the twenty-eight who escaped, sixteen speak here, either directly or through the other survivors with whom they traveled. These men survived not only the clumsily orchestrated massacre itself, but many days in the wilderness wounded and without food, water or comfort. Several were recaptured by the enemy and managed to escape and survive yet another round.

Some men traveled alone to try to find the Texian army. Others met up en route and traveled in groups of threes or fours. The survivor accounts here are presented in the chronological order in which they appeared, from William Hadden in April of 1836 to John C. Duval in 1867.

What this volume is *not* intended to be is a military history of the Battles of Coleto, San Patricio, Agua Dulce, Refugio or other events that are historically conjoined to the massacre. Scrutiny of the leadership of Colonel Fannin has likewise been left to military scholars who have tackled the subject in a capable fashion. The Battle of Coleto was left intact in some survivor accounts to preserve the continuity of the narratives, or, in the case of Captain Jack Shackelford, to provide an officer's view of the events, since Fannin is necessarily silent on the matter.

The informed reader will note the absence of a narrative by one of Goliad's best-known survivors, Herman Ehrenberg. He published, in several editions, a lengthy account of his experiences in the Texas Revolution, including his brush with death at Goliad.

Unfortunately, there is no single English translation to date that I am comfortable presenting here. Though it has been translated from the German on three known occasions, there is still want of a reliable, scholarly translation. Such a feat will require a translator who has no biases, and who has both the language credentials and historical knowledge of the Revolution to render Ehrenberg's words correctly and justly.

It is still known in some circles today that the rallying cry at the Battle of San Jacinto was "Remember the Alamo! Remember Goliad!" Folks from here to Japan know that they're supposed to remember the Alamo and a few will even know why, but rare is the soul who remembers La Bahia. Why?

Both the storied mission and the historical presidio (no, La Bahia is *not* a mission contrary to popular belief) represent Texian military losses. Both have elements of surprise woven into their missives. But the defenders of the Alamo had a thirteen day siege to contend with, during which Travis penned a handful of letters begging for assistance and declaring "Victory or Death!" Fannin, on the other hand, had no opportunity to send messengers. Travis and his men in the Alamo had Santa Anna to deal with directly, to demand that they surrender at discretion. Fannin had subalterns to reckon with, who tried to play by the standard rules of warfare until Santa Anna ordered them to commit mass murder. The Alamo had within its walls "rock stars" of the day…David Crockett and James Bowie. Goliad had naught. So, the Alamo siege can be romanticized and wrapped up neatly in a John Wayne movie. Goliad, on the other hand is too complicated and messy for the movie-going mind. What with the battle, surrender and captivity, it would take several installments of a mini-series to get to the treacherous surprise ending. And besides, John Wayne is dead.

What the Alamo truly did for Texas, other than spawn a bunch of legends and indeed some bona fide heroes, was to keep Santa Anna's hands busy while the Declaration of Independence was being signed and a constitution was being drawn up. Nobody had any way to know that His Serene Highness would allow the siege to continue for so long before storming the fort, but so long as he was in Bexar and not in Washington, a nation could be born. What Goliad did for Texas was give our fighting men, particularly at San Jacinto, a lust for vengeance and a the knowledge that the options truly *were* victory or death. Goliad had proven that Santa Anna was willing to spill the blood of as many Mexican soldiers as it took to win, and intended to kill every colonist as a traitor and everyone else as a pirate. Goliad gave the men at San Jacinto cold-blooded, ruthless slaughter and 342 more souls to avenge.

The location of the Alamo, in tourist-infested downtown San Antonio, of course, gives it a leg up as far as visitor stats are concerned. What remains of it bears little resemblance to what it looked like when it fell on March 6, 1836. But tourists, foreign and domestic, don't much care. After Sea World, overpriced enchiladas and snow cones, it sure feels good to say you visited an historical place, doesn't it?

To get to Presidio La Bahia, on the other hand, unless you're fortunate enough to live nearby, you have to *intend* to go there. You have to veer a little out of the way. But when you're there, you can see the chapel, the bastions—in short, a true fort. You are afforded the benefit of a beautiful and complete stone armature to wrap your imagination around. The Alamo lacks this. San Jacinto, of course, never had it. Goliad does. Wander the grounds at your leisure or sit in repose in the serenity of the chapel. Stand on the bastion where Abel Morgan stood guard on the night of March 19, 1836 or from whence Captain Jack Shackelford watched Horton's men skirmish that day. Walk the earth where the wounded prisoners were hauled out and shot. You can even get enchiladas and snow cones afterward.

Lest it might be misconstrued that I dislike the Alamo, I should say adamantly and with fervor that I do not. On the contrary, I appreciate that *some* part of the Texas Revolution is known the world over. That's an honor that is uniquely Texan. It's a source of pride. The Alamo should be a sacred place in the heart of every Texan, but no more so than Goliad or the battleground at San Jacinto. So the next time someone discovers you're a Texan and makes that inevitable Alamo reference, consider it your duty to remember Goliad to yourself and to your new acquaintance. Bring the kids or the grandkids down for a day to walk in the footsteps of those who perished and those who undertook remarkable adventures to preserve their lives. If we forget to remember La Bahia, we've forgotten about half of the men who died in the name of Texas.

To Newton Warzecha, director of Presidio La Bahia, this book is dedicated with respect and admiration, in tribute to your enthusiasm, plain-spoken nature, and ferocious defense of history and the garrison in your charge.

-Michelle M. Haas, Managing Editor
Windy Hill

Semantics of a Massacre

Most survivor accounts put forth that Colonel Fannin did not surrender at discretion; that he and General Urrea agreed upon terms of capitulation, which included the Texians being granted prisoner of war status, respect for their personal property and parole to the United States within a week. It is noted in most accounts that these terms were written in both English and Spanish, and signed by officers on both sides. The only extant copy remains in Mexico City to which Urrea later appended a note stating that he did not grant Fannin any terms and that the surrender was unconditional. Wasn't the purpose of the surrender to *avoid* wholesale bloodshed on both sides and to preserve the lives of the wounded? If the men had any notion that their options were to die fighting or be shot as pirates or traitors, would they have consented to a surrender, particularly as the Alamo still smoldered? Colonel Fannin must have had enough assurance from General Urrea to satisfy his mind. We, of course, cannot know.

Urrea moved on to Victoria, leaving Lieutenant Colonel Portilla in charge of Goliad. Urrea sent word to Santa Anna of the surrender and prisoners, recommending clemency for Fannin and his men, and Santa Anna responded by signing their death warrants. He invoked the Tornel Decree and was very upset that the prisoners had *not yet* been executed. Officers and men in his army, however, reported having a difficult time complying with orders so alien to the rules they knew as soldiers. Portilla, on the day of the massacre, informed Urrea that he would remain a dutiful officer, but begged that his future duties not include carrying out another Goliad.

The bloody decree was passed December 30, 1835. It was published and distributed in Mexico in January of 1836. Its first known publication in the United States was at New Orleans on

February 13, 1836, and it did not reach Texas newspapers until after the fall of the Alamo in March. Santa Anna used these two short paragraphs to justify taking the lives of most of the prisoners at Goliad:

1) Foreigners landing on the coast of the republic or invading its territory by land, armed with the intention of attacking our country, will be deemed pirates and dealt with as such, being citizens of no nation presently at war with the republic, and fighting under no recognized flag.

2) All foreigners who will import either by sea or land, in places occupied by the rebels, either arms or ammunition of any kind for the use of them, will be deemed pirates and punished as such.

That's the Tornel Decree…the inviolate supreme law of the land ruled by a dictator who would enforce it (or not) at his whim. He could add to or subtract from its meaning as he saw fit on any given day. And he did. Seems it wasn't so inviolate after all. After his defeat at San Jacinto and after being deposed from his dictatorial perch, Santa Anna tried to account for his actions in his 1837 *Manifesto*. "How my arbitrariness would have been exaggerated," he cooed, "and to speak the truth with justifiable excuse, if by pardoning, as I desired, those unfortunate wretches I should have dared to violate the law." As Urrea and the army knew, Santa Anna *was* the law in 1836.

The Tornel Decree found its voice in circumstances where its maritime language was better suited—armed Americans landing on Mexico's shores. Shortly after Santa Anna squashed the constitution and the congress in late 1835, an anti-Centralist expedition was arranged by the exiled Mexican general, Jose Antonio Mexia, to take Tampico and spread anti-Centralist sentiments further down the eastern coast of Mexico. He raised volunteers at New Orleans with land as his lure and they sailed November 6, 1835. Most everything went wrong upon arrival. Mexia was defeated and withdrew, leaving behind thirty-one prisoners. Three died from injuries, while the remaining twenty-eight were tried by a military court for *piracy*. They were convicted and sentenced to be executed on December 14, 1835. The opinion in the United States seemed to be that Mexico was well within her rights to carry out those sentences. And in less than two weeks, the Tornel Decree was born—piracy, foreigners and all.

Were there trials for the "pirates" of Goliad? Pirates, like other suspected criminals, were tried for their crimes before being executed, as Santa Anna himself demonstrated at Tampico. At Goliad, there were none. At very least, were interviews ordered for those captured to determine who the *foreigners* were? No. For the men of Goliad, Santa Anna acted alone as judge and jury but was too busy to personally act as executioner. Those slaughtered under the Tornel Decree, which expressly applied to foreigners, were both colonists and Americans.

William Hadden, for example, was a resident of Texas after the Declaration of Independence was signed...before that document, he had been a legal resident of Mexico. In Santa Anna's eyes, there was no separate republic. It was *all* Mexico. Hadden was still a Mexican *citizen* to His Excellency. So William Hadden should have been spared the "pirate treatment." He wasn't. And General Urrea knew from experience that citizenship didn't matter to his boss.

Just a few weeks prior to the surrender at Goliad, General Urrea inquired of Santa Anna what he was to do with prisoners taken at San Patricio who were not foreigners, but who were Mexican citizens. Santa Anna replied, in the midst of the Alamo siege, on March 3:

> ...since, in my estimation, the Mexican who engages in the traitorous act of joining these adventurers, as described, loses his rights as citizen according to our laws, the five prisoners to whom your Excellency refers should also be treated likewise.

The fate of Captain Miller's men of the Nashville Battalion, provides another example of arbitrary interpretation of the Tornel Decree by Santa Anna. The following, written by Ramon Caro, Santa Anna's secretary at the time, addresses how it came to pass that Miller's men were not victims of the massacre. It seems that a large body of men who were not citizens of Mexico, who were en route to Goliad to assist Fannin and who had arms in their possession for that purpose would have been choice cuts for butchery under the Tornel Decree. But regardless of where they came from or where they were going, they would have been put to death anyway had it not been for one Mexican officer's willingness to plead to the contrary. This allowed Miller's men the benefit of an investigation—something that the other "pirates" had not the benefit of. Since the investigation was yet in progress on the 27th, Miller's men escaped being mowed down near the Presidio. Instead, they were taken to Matamoros and imprisoned.

From Ramon Caro's 1837 "True Account of the First Texas Campaign," as found in *The Mexican Side of the Texas Revolution*:

> ...The commandant of Bahia notified His Excellency at about the same time that 83 men had been taken prisoners by Colonel Vara. He sent him the original report of Colonel Vara in which it appeared that five men who were making their way to the fort, ignorant of the surrender of Fannin, declared that they had just landed at Copano and that their companions were still on board the vessel that had brought them. The colonel informed them of Fannin's surrender and told them to ask their companions to land and surrender, promising them that they would be treated with all consideration if they surrendered without offering resistance. They acceded to his request and were all taken to the fort to await the disposition of His Excellency. When he received this information, he ordered me to write to the commandant, giving him instructions to have all the prisoners executed, as provided by the circular of the supreme government, for though they had not engaged in active fighting, the fact that they had introduced themselves into the country armed, confirmed their intention of taking an active part in the war.

> Fortunately, when Captain Savariego, bearer of the order, learned that it extended to the eighty-three men, he asked to be allowed to speak to His Excellency, and I myself led him to the room where he was. Captain Savariego told him that the colonel who had taken these men had asked him to ask the clemency of His Excellency, for the unfortunate prisoners who had surrendered without making use of their arms. Hardly had he spoken, when for his reply he received such bitter reproof that he left the room disgusted.
>
> At the same time, His Excellency called me and ordered me to alter the order which had already been written in final form, instructing the commandant of Bahia to hold the eighty-three prisoners until a thorough investigation was concluded concerning the circumstances of the surrender, allowing them only one ration of meat a day. The investigation was immediately instituted by General Cos, who used Lieutenant Colonel Pedro Francisco Delgado, a member of the secretarial staff of His Excellency, as his secretary. He took the declaration of Captain Savariego and sent it to Bahia from where it was to be returned as soon as possible with the additional information necessary for a final decision of the case. Those unfortunate wretches escaped a tragic end by this coincidence, for they would have been executed in spite of the fact that such an act was contrary to the spirit of the circular of the supreme government.

There are those who today continue to beat the war drum around La Bahia, 177 years after the bodies of the slaughtered prisoners were tossed in heaps to roast, and the few who escaped fled, choking on the smoke of the burning corpses of their comrades. There is no language barrier as there was inside the fort in March of 1836…both sides have ably given their viewpoints in plain English to the media. One side is driven by race; the other by history. One side is concerned with semantics and emotions; the other with history.

At the heart of the debate are worries over words. Some feel that the word *massacre* should not be associated with Goliad, since what took place there was a *lawful execution*. Some are concerned that *massacre* is an insensitive word and somehow reflects badly on the Mexican people of the day and of today. Rather than looking critically at what lead to the pulling of so many Mexicans triggers March 27, 1836, these individuals and groups are more troubled with how certain words make them *feel*. We cannot alter the events of 177 years ago in order to pacify delicate sensibilities or injured feelings today. The words that were, and are, used to describe what happened at Goliad had very particular meanings assigned to them in 1836 and those definitions have not changed much, regardless of how we may feel about them.

The question of race, so often brought to this debate, in this instance is a non-issue. History has, and must continue to, paint a portrait of Santa Anna as a treacherous, narcissistic and dangerous man. An indictment of Santa Anna is not any kind of commentary on Tejanos or Mexicans, then or now. Any discussion of race is merely a tool with which to gain leverage in an historical debate where the facts are abundantly clear, if not to their liking: one very bad Mexican despot is responsible for ending the lives of nearly 350 men on March 27, 1836.

Anyone who wishes to rewrite the history of the Presidio based solely on the offending words, massacre and execution, should first consult a dictionary. The first edition of Webster's fine reference (1828) would be best, but any dictionary will do. On the subject of massacres, Mr. Webster was very precise. From Webster's 1828:

massacre

n. 1. The murder of an individual, or the slaughter of numbers of human beings, with circumstances of cruelty; the indiscriminate killing of human beings, without authority or necessity, and without forms civil or military. It differs from assassination, which is a private killing. It differs from carnage, which is rather the effect of slaughter than slaughter itself, and is applied to the authorized destruction of men in battle. Massacre is sometimes called butchery, from its resemblance to the killing of cattle. If a soldier kills a man in battle in his own defense, it is a lawful act; it is killing, and it is slaughter, but it is not a massacre. Whereas, if a soldier kills an enemy after he has surrendered, it is a massacre, a killing without necessity, often without authority, contrary to the usages of nations, and of course with cruelty. The practice of killing prisoners, even when authorized by the commander, is properly massacre; as the authority given proceeds from cruelty.

execution

n. …4. The last act of the law in the punishment of criminals; capital punishment; death inflicted according to the forms of law.

Essentially, folks who are massacred are defenseless and often taken by surprise, whereas the executed are folks who have been sentenced for a crime. Not a soul that was fired upon at Goliad had been tried, convicted or sentenced. All those killed en masse without prior knowledge that death was imminent had none of the opportunities afforded to convicted criminals to get their mental, physical or spiritual affairs in order. They were led out under false pretenses and taken down like fattened beeves ripe for slaughter. That is the description given by both the survivors and those on the Mexican side who, likewise, referred to the event as a massacre in description and by name.

While the deposed Santa Anna would have readers believe that he wept for every life that was lost at Goliad, Santa Anna while in power expressed indignation that everyone in and around Goliad, including Miller's men captured at Copano, were not shot *immediately upon surrender*. The men who were shot believed that the conventions of war of civilized nations applied to that which they were fighting; the men who shot them believed the same. History, in this instance, was not just written by the victors.

Thanks to the survivors, *both* sides told the story. And both sides told the same story of the events of that morning—the story of a massacre. The only person who considered the massacre a *lawful execution* was Santa Anna himself.

One sporadic dictator's whims and a decree created to speed his victory, arbitrarily enforced according to those whims did not change the accepted conventions of war as they were at the time.

The man who orchestrated the events of the 27th of March 1836 did not even attempt to present it as an execution. He did not lift a finger (or pull the strings to lift a finger of one of his minions) to put a pretty face on his ghoulish creation. Nor should we. To call the Goliad Massacre anything but is to dishonor those who stood and faced immediate death on that Sunday, and to insult those subalterns of whom Santa Anna would settle for no less than cold-blooded murder in the name of power.

Urrea's Lament

[Excerpts from the Diary of Military Operations under the command of General Jose Urrea, excerpted from Casteñeda's The Mexican Side of the Texas Revolution.*]*

...At half past six in the morning the ammunition arrived which, as stated before, had been lost the day before; and although more had been ordered from Col. Garay, this had not arrived up to this time. One hundred infantry, two four-pounders and a howitzer were added to my force. I placed these as a battery about 160 paces from the enemy protected by the rifle companies. I ordered the rest of the infantry to form a column that was to advance along the left of our battery when it opened fire.

As soon as we did this and began our movement as planned, the enemy, without answering our fire, raised a white flag. I immediately ordered my battery to cease firing and instructed Lieut. Col. Morales, Captain Juan José Holzinger, and my aide, José de la Luz González to approach the enemy and ascertain their purpose. The first of these returned soon after, stating that they wished to capitulate. My reply restricted itself to stating that I could not accept any terms except an unconditional surrender. Messrs. Morales and Salas proceeded to tell this to the commissioners of the enemy who had already come out from their trenches. Several communications passed between us; and, desirous of putting an end to the negotiations, I went over to the enemy's camp and explained to their leader the impossibility in which I found myself of granting other terms than an unconditional surrender as proposed, in view of which fact I refused to subscribe to the capitulation submitted consisting of three articles.

Addressing myself to Fannin and his companions in the presence of Messrs. Morales, Salas, Holzinger and others I said conclusively, "If you gentlemen wish to surrender at discretion, the matter is ended, otherwise I shall return to my camp and renew the attack." In spite of the regret I felt in making such a reply, and in spite of my great desire of offering them guarantees as humanity dictated, this was beyond my authority. Had I been in a position to do so, I would have at least guaranteed them their life. Fannin was a gentleman, a man of courage, a quality which makes us soldiers esteem each other mutually. His manners captivated my affection, and if it had been in my hand to save him, together with his companions, I would have gladly done so. All I could do was to offer him to use my influence with the general-in-chief, which I did from the Guadalupe.

After my ultimatum, the leaders of the enemy had a conference among themselves and the result of the conference was their surrender according to the terms I proposed. They immediately ordered their troops to come out of their entrenchments and to assume parade formation. Nine pieces of artillery, three flags, more than a thousand rifles, many good pistols, guns, daggers, lots of ammunition, several wagons, and about 400 prisoners fell into the hands of our troops. There were ninety-seven wounded, Fannin and several other leaders among them. I gave orders for all the baggage to be taken up, the prisoners to be escorted to Goliad by 200 infantry, and the wounded who were unable to walk to be carried in the carts or wagons taken from the enemy. They had lost twenty-seven killed the day before. I lost eleven killed, and forty-nine soldiers and five officers were wounded. Capt. José María Ballesteros was seriously injured.

Right on the battlefield, I wrote a note to Col. Garay telling him of the outcome and asking him to make a report to the general-in-chief, for it was impossible for me to do it at the time because I was marching to Guadalupe Victoria without stopping to rest. Through a dispatch from Col. Garay, I learned that he had taken possession of Goliad where he had found eight pieces of artillery which the enemy had been unable to take with them. When he took possession, the houses of the city were still burning, having been set on fire by the enemy before it retreated. Combustible materials were left to prolong the fire, and very few houses were saved…

Urrea's Lament

...I spent the 24th, 25th, 26th, and 27th in organizing my forces, equipment, and ammunition, and in drawing up many instructions for the security of the military posts that I was leaving on our rear, as well as for the better care of the wounded who have been up to now in the hands of a bad surgeon. As among the prisoners there were men skilled in all trades, I secured surgeons from among them who were very useful to us as well as to the sick in the hospital at Bexar, where I sent those that were needed.

On the 25th I sent Ward and his companions to Goliad. The active battalion of Querétaro joined me on the 26th. On the 27th, between nine and ten in the morning, I received a communication from Lieut. Col. Portilla, military commandant of that point, telling me that he had received orders from His Excellency, the general-in-chief, to shoot all the prisoners and that he was making preparations to fulfill the order.

This order was received by Portilla at seven in the evening of the 26th, and although he notified me of the fact on that same date, his communication did not reach me until after the execution had been carried out. All the members of my division were distressed to hear this news, and I no less, being as sensitive as my companions who will bear testimony of my excessive grief. Let a single one of them deny this fact! More than 150 prisoners who were with me escaped this terrible fate; also those who surrendered at Copano and the surgeons and hospital attendants were spared. Those which I kept, were very useful to me as sappers.

I have come to an incident that has attracted the attention of foreigners and nationals more than any other and for which there have not been lacking those who would hold me responsible, although my conduct in the affair was straightforward and unequivocal. The orders of the general-in-chief with regard to fate decreed for prisoners were very emphatic.

These orders always seemed to me harsh, but they were the inevitable result of the barbarous and inhuman decree which declared outlaws those whom it wished to convert into citizens of the republic. Strange inconsistency in keeping with the confusion that characterized the times! I wished to elude these orders as far as possible without compromising my personal responsibility; and, with this object view, I issued several orders to Lieut. Col. Portilla, instructing him to use the prisoners for the rebuilding of Goliad. From time on, I decided to increase the number of the prisoners there in the hope that their very number would save them, for I never thought that the horrible spectacle of that massacre could take place in cold blood and without immediate urgency, a deed prescribed by the laws of war and condemned by the civilization our country. It was painful to me, also, that so many brave men should thus be sacrificed, particularly the much esteemed a fearless Fannin. They doubtlessly surrendered confident that Mexican generosity would not make their surrender useless, for under any other circumstances they would have sold their lives dearly, fighting to the last. I had due regard for the motives that induced them to surrender, and for this reason I used my influence with the general-in-chief to save them, if possible, from being butchered, particularly Fannin. I obtained from His Excellency only a severe reply, repeating his previous order, doubtlessly dictated by cruel necessity. Fearing, no doubt, that I might compromise him with my disobedience and expose him to the accusations of his enemies, he transmitted his instructions directly to the commandant at Goliad, inserting a copy of order to me. What was done by the commandant is told in his diary. Here, as well as in his communications, are seen the motives that made him act and the anguish which the situation caused him. Even after this lamentable event, I still received a letter of the general-in-chief, dated on the 26th, saying: "I say nothing regarding the prisoners, for I have already stated what their fate shall be when taken with arms in their hands."

In view of the facts presented, and keeping in mind that while that tragic scene was being enacted in Goliad I was in Guadalupe Victoria, where I received news of it several hours after the execution, what could I do to prevent it, especially if the orders were transmitted directly to that place? This is to demand the impossible, and had I been in a position to disregard the order it would have been a violent act of insubordination. If they wish to argue that it was in my hand to have guaranteed the lives of those unfortunates by granting them a capitulation when they surrendered at Perdido, I will reply that it was not in my power to do it, that it was not honorable, either to arms of the nation or to myself, to have done so. Had I granted them terms, I would then have laid myself open to a trial a council of war, for my force being superior to that of the enemy on the 20th and my position more advantageous, I could not admit any proposals except a surrender at discretion, my duty being to continue fighting, leaving the outcome to fate. I belie that I acted in accordance with my duty and I could not do otherwise. Those who assert that I offered guarantees to those who surrendered, speak without knowledge of the facts.

[Urrea included a copy of the capitulation, below, in the 1838 publication of his Diarios, *published party in answer to Santa Anna's* Manifesto. *The text appended below the terms of capitulation, effectively nullifying them, were obviously added by Urrea after Fannin and his men signed the document. It is undated and could have been added moments after the document was signed, or long after the Massacre. It also may well have only been added to the Spanish copy and not the English translation that was given to Fannin.]*

Art. 1. The Mexican troops having placed their battery at a distance of one hundred and seventy paces from us and the fire having been renewed, we raised a white flag. Colonels Juan Morales, Colonel Mariano Salas and Lieutenant Colonel Juan José Holsinger of Engineers came immediately. We proposed to them to surrender at discretion and they agreed.

Art. 2. The commandant Fannin and the wounded shall be treated with all possible consideration upon the surrender of all their arms.

Art. 3. The whole detachment shall be treated as prisoners of war and shall be subject to the disposition of the Supreme Government.

Camp on the Coleto between Guadalupe and La Bahia, March 20, 1836. B. C. Wallace, commandant, J. M. Chadwick, Aide.—Approved, James W. Fannin.

Since, when the white flag was raised by the enemy, I made it known to their officer that I could not grant any other terms than unconditional surrender and they agreed to it through the officers expressed, those who subscribe the surrender have no right to any other terms. They have been informed of this fact and they are agreed. I ought not, nor can I grant any other terms—José Urrea.

[The above is a translation of the Spanish language copy of the capitulation agreement in the Archives of War Department in Mexico City.]

Hadden in Harrisburg

[Though William Hadden left no written account of his escape from the Goliad Massacre, a letter in the papers of the Texas Revolution references his escape and safe his arrival at Harrisburg. It is unknown to whom the letter was addressed or who wrote it. Likely, it was written to the editors of a New Orleans newspaper for publication there. It was carried in the Morning Courier & New York Enquirer *of May 14, 1836. The letter is dated April 7, just a few days after Sam Houston himself received the sad news and wrote to Secretary of War Rusk about it. Both Houston and the author of this letter believe, at the time of writing, that there had been but one survivor of the Massacre. Such was the way news traveled in a land as big as Texas and a land that was in tumult. William Hadden had long been a friend of Texas. He came with his family to Texas as a boy, around 1823, and had given his signature to the Goliad draft of the Declaration of Independence under Dimmit. He was among those of A. C. Horton's Mounted Rangers who, likely for lack of a mount, fought in the Battle of Coleto and were imprisoned after the surrender. This appears to be the earliest survivor account following the Massacre on the 27th of March. The letter can be found in Volume 5 of Jenkins,* Papers of the Texas Revolution.*]*

Harrisburg,
7th April, 1836

Dear Sir—This place is now the seat of government of Texas. These are exciting times; every man now walks with a full heart—moistened eyes, and compressed lips—indicating his commingled feelings of grief and anger. Young Hadden is here; he alone escaped, of all the forces which were with Col. Fanning. On the 19th ult., Col. Fanning, who had weakened his forces by sending parties on various expeditions, commenced his retreat from Goliad, having with him about three hundred men. Two or three leagues on this road, he was attacked by a Mexican army of near three times his numbers—infantry, cavalry and artillery. It was two hours before night, and the Texians defended themselves with the most undaunted resolution against the repeated charges and most vigorous efforts of the enemy, until dark, when they drew off their forces—having lost in killed and wounded 180 men. The Texians had ten men killed and several wounded. Among the latter was Col. Fanning himself. During the night the Texians entrenched themselves in their position on the prairie. The next morning they found themselves surrounded by the Mexican army, who sent an officer with a white flag and he was met by another from the Texians and a capitulation was entered into, by which it was stipulated that the Texians should surrender as prisoners of war—that they should retain their private property, and that they should be sent to New Orleans, and be released upon their parole of honor not to serve again against the Mexicans during the war.

The reasons which induced the surrender without making further defence was the want of water, and a belief that the Mexicans would comply with the articles of capitulation, according to the custom of all civilized nations.

The prisoners were all marched back to Goliad, and were joined by other parties of prisoners, amounting in all to 407 men. They were kept eight days in confinement—being allowed only a small piece of beef to each man once a day, and no bread; on the 9th day at sunrise, they were marched out under the pretence of taking them to Copano, to embark for New Orleans. The prisoners rejoiced at the prospect of a speedy release—but when they found themselves divided into four parties, each with a guard of overwhelming strength, and that these parties were marched by different routes—they began to have gloomy forebodings. When they had marched about a quarter or half a mile, young Hadden's company heard a firing in the direction of Col. Fanning's party. A murmur arose that they were killing the prisoners; at that moment the guard, which were in two lines, one upon each side of the prisoners, passed all to one side and commenced a fire by platoons at the prisoners. Young Hadden and three others yet unhurt, started and fled. The cavalry, armed with lances, pursued them; they plunged into the river and swam—one was killed in the water, one upon the bank, and the fate of the third is unknown. Young Hadden secreted himself until night. From his hiding place he heard the shrieks and groans of the wounded and dying men, the cries of "O Lord! O Lord have mercy!" The most affecting ejaculations of distress were mingled with the noise of guns which gradually subsided towards midday, when the horrid work was finished.

The neighbors who have known Hadden from his childhood say that his statement may be relied upon with implicit confidence; no man can hear him tell his story and doubt its truth, and a blacker picture of perfidy and blood is not on record in the annals of history.

[Hadden surfaces again in 1836, this time imparting information received from a chat with Susanna Dickinson about the fall of the Alamo to the father of one who fell there. William Parker mentions Hadden in writing to the Mississippi Free Trader *on April 29, 1836, excerpted below.]*

To the Editor of the Free Trader:

Sir,—Having just returned from a trip to Nacogdoches, made for the purpose of obtaining information concerning the fate of my son, who feel in defending the Alamo, I take this method to correct the erroneous news in circulation among my fellow citizens concerning the position of the belligerents in Texas, and concerning the part which the Indian tribes, near the frontiers, are likely to act in this momentous and exterminating crusade against civil and religious liberty.

...The hard and unprecedented fate of Col. Fannin and his brave companions is but too well confirmed, by the latest advices at headquarters, said my informant, to admit any longer even a hope that they still live. After stipulating, at the time of surrender, for the lives, subsistence and transportation of himself and men on a Mexican vessel from Copano to New Orleans...this champion of liberty and surviving heroes, who but nine days before had sustained an action of nine hours duration in the open field against 8 to 10 times their numbers, after marching a few miles from Goliad in the direction of Copano, were fired upon by the detachment sent to conduct them to the vessel, and all killed but one. A private, by the name of William Hadden, and three others, escaped the first fire and ran off. Hadden fell in very high grass, and behind a bush—and perceiving that he was not pursued, lay concealed until night—the other three were shot at the second fire by their pursuers. After sunset, and after the assassins had plundered the dead and retired to the fort, Hadden made his escape, crossed the San Antonio River, and reached the Colorado on the 8th day, when he was accidentally found in a state of starvation by Major Benjamin J. White, and brought into the settlements on the Brazos. It is upon Hadden's statement that these details are founded—And but for his escape, in a manner most miraculous, the world must have remained ignorant for a long time to come, perhaps forever, of this atrocious and cold blooded massacre by the enemy, and of the unrevenged and yet unpunished martyrdom of the volunteer heroes of the Coleto...

-William Parker
Natchez, April 29th, 1836

Joseph Field - Post Surgeon

[The following account of Dr. Field's experiences at La Bahia is taken from his book, Three Years in Texas, *published in Massachusetts in September 1836. He had officially tendered his resignation as Post Surgeon for Copano in December of that year, after which he moved about Texas practicing medicine. In the 1870s, he applied to the state for his pension, at one point writing from Rockport en route to Mexico, hoping he could perhaps live in more comfort there on the little means available to him. By 1878, he had taken up residence in Hillsborough County, Florida, where his pension payments were sent, and where he died in 1882.]*

...On the night of the sixth of March, I arrived at La Bahia de Goliad, having travelled seventy miles alone, through a country so beset with Indians, as to make it necessary to ride by night. Here I found Col. Fanning, to whom I reported myself, he having, at the approach of the enemy, fallen back to the fort at the above place.

[After learning the fate of Ward and King] a council of war was immediately held, at which it was determined to execute the General's order [to retreat and join the main army] as speedily as possible. The Council was hardly dissolved before our spies, who had been sent out on the road to San Antonio, returned in great haste, saying that the Mexican army was within six miles of us. The order was instantly changed, and active preparations were made for the defence of the fort. Our precautions for defence were not hindered until the next day, when there was some skirmishing on the other side of the river between our cavalry and theirs, without loss to either. Believing their numbers not to exceed five hundred, and feeling ourselves able to cope with that number, though we were not then more than about three hundred, Col. Fanning made arrangements to decamp at evening of the same day. When it was dark—and it was very dark—Capt.

Horton, with his company of cavalry, being sent to occupy the ford of the river, one mile from the fort, returned with information, that a body of troops were on the opposite bank, and that they attempted to charge upon him. His opinion that the retreat should be delayed until morning, was adopted.

At an early hour the next day we were under marching orders. Our cannon, baggage and sick, were drawn by Mexican oxen, in Mexican carts. Not being well broke, nor understanding the language and manners of English drivers, many of them as they issued from the fort, ran furiously into the prairie, and were unmanageable. Others would go no way but backwards. We were in the midst of a wide prairie, when suddenly, the enemy were seen displaying their columns two or three miles in our rear. Our commander ordered a halt, unlimbered the cannon, and forming a line fronting the enemy, commenced cannonading them. We had not made many shots before the order was to fasten the oxen, all of which had been turned loose, to the carts. The line of battle was ordered to file to the rear from the extreme right and left wings, marching in double columns, our baggage in the centre, and cannon in the rear. In this manner we continued retreating, until being surrounded, we were compelled to stop and fight.

Our situation was very unfortunate, being in the midst of that large prairie, in a place where the ground was much lower than that around us. We were also without water, which is the greatest of necessities, especially for the wounded. The enemy having closed around us upon every side made a general charge but were repulsed with great slaughter. They rallied and charged again and again; but at every succeeding charge, with less vigor, until night came and put an end to the carnage. The enemy retired to the woods in the direction of our march. When they had taken their position for the night, Col. Fanning ordered his men to prepare for resuming their march and cutting their way through the enemy's lines. But it was soon discovered that so many of our horses were killed or wounded, and our oxen strayed away, that it was impossible to transport our wounded, who were more than sixty in number. Our commander said he would not leave them, but was resolved to share with them a common fate.

I will not attempt to describe the horrors of that night, which was spent amid the groans of the dying and the incessant cry for water of the wounded, in digging ditches and erecting breastworks for an expected engagement on the following day. Morning at length came, and with it came our enemy, marshalling themselves in battle array, with apparent intention of renewing the scenes of the previous day, when a cannon ball was fired over us from near the woods. It was twice repeated, when almost simultaneously, a white flag was raised upon both sides. When the two commanders met at a proper distance from their respective armies, the Mexican General Urrea embraced Colonel Fanning and said, "Yesterday we fought; but today we are friends."

Articles of capitulation were soon agreed upon by the two commanders, and committed to writing with the necessary signatures and formalities. The articles were, that in consideration of our surrendering, our lives should be ensured, our personal property restored, and we were to be treated, in all respects, as prisoners of war are treated among enlightened nations. We also received a verbal promise to be sent, in eight days, to the nearest port, to be transported to the United States.

The main body of the prisoners were marched back to the fort at Goliad in the afternoon of the same day but the Colonel, Dr. Barnard and myself, encamped upon the ground.

In the morning, Colonel Holsinger, who was left in command of the guard, inquired of the Colonel which was his best surgeon. As I was standing near, he pointed towards me, and said he believed I was as good as any. I was then ordered to follow a carriage, in which were placed two wounded Mexican officers, whom I followed on foot. When I entered the fort, I was taken to the church, where our prisoners were confined, and thrust in among them; though there was not room for more than two-thirds to sit at the same time. Here our only resting place was the bare ground, offensive with filth; Mexican churches being without floors, or any marks of cleanliness. I was soon called for by Col. Holsinger, and put in a house outside the walls of the fort, with the wounded Mexican officers, in whose company I left the battlefield. By apparent sympathy and good attention to their wounds, I obtained from them expressions of friendship and confidence.

They often spoke to me of going with them to Mexico and living with them, not as a prisoner, but as a friend. Relying a little on these testimonies of friendship, I requested that my friend, Capt. John S. Brooks, who had broken his thigh in the engagement, might be brought into the same room, that I might pay more attention to his wound. The request was granted and he, when he arrived, after having lain in an open cart in the prairie without food or other refreshment for three days, was permitted to be placed by the door, upon the hard ground.

I also ventured to ask that a servant might be sent into the prairie to collect some grass to make a bed for him; and though I offered all my money, which amounted to more than one hundred dollars, which I happened not to have about me when I was robbed, it was not granted. They knew a much easier way of getting my money, which they afterward obtained. Our whole time (the surgeons) was taken up in dressing the Mexican wounded. All our medicines, surgical instruments and bandages were taken from us; and none of our soldiers had their wounds dressed, except a few by Major Miller. Our treatment during the week was like that of which I have given a specimen.

In the night of the 27th [26th] the express returned, who had been sent to Bexar to know the will of the President, Santa Anna, with an order for all prisoners to be shot.

About sunrise the next morning, the sanguinary mandate was executed. Being outside of the fort I was an eyewitness to but a small part, but was informed of what was going on. At length, an officer and two soldiers came in, one of whom seized the blanket which covered Capt. Brooks, which I indignantly pulled from him and replaced. Seeing it, my friend, the wounded Mexican

Captain called me and made me take a seat by him upon the bedside and hold my tongue. When they rudely bore away Capt. Brooks, he extended his arms towards me, imploring my assistance, until his voice was silenced forever. In a short time afterward, I was taken by a subaltern to a house where I found Mr. Spohn, and where I received some food. When we left this house, we were taken to the hospital.

When the confusion of the morning had a little subsided, a division of the wounded was made, and an equal number given to each surgeon to attend. We were then crowded into an apartment too small to allow all to lie down at once with convenience. Some of us were so fortunate as to have a blanket, which we spread over the filthy ground to sleep upon. Our food was chiefly stewed beef, which we ate from our fingers. In a week or two we had a small allowance of bread.

During my captivity, I embraced every opportunity to make myself acceptable to the wounded Captain—listening, with seeming interest, to whatever he had to say of his own and my future happiness in the Mexican country, and manifesting impatience at the unavoidable delay caused by the obstinacy of his wounds.

In the meantime all my thoughts were employed in contriving means of escape. As I could not so well go alone, I proposed to such of my fellow prisoners as I thought worthy of confidence, consulting one at a time, to make a trial at elopement. Several weeks elapsed before I could find one whose prudence justified a trial. They all believed it impracticable and that a failure would be attended with immediate death.

At length a German by the name of Vose, whose impatience under repeated insults had subjected him to many mortifying punishments, came to an understanding with me. The time, manner and place of meeting having been agreed upon, we by various pretexts, obtained permission to sleep outside of the walls. When it was dark I took a path leading to the river, where we commonly went for water, appearing to go for that purpose, and having descended to the brink, I shaped my course up the river under a steep bank and projecting rocks. Climbing precipices, which, under other circumstances would have been insurmountable, I at length gained the plain and place of meeting, where I found my companion waiting for me.

We then ascended the San Antonio River about one mile, where we found a place that was fordable. Having crossed, we descended an Indian trail leading to the north, the same that I came in upon a few weeks before. Our course led us in the direction of the Guadalupe River, where we arrived the following day. Here for the first time, my companion informed me of his inability to swim. Setting my ingenuity to work, I soon constructed a raft of rails and other trash that I found upon the bank, sufficiently large to float him across, and making a line fast to it, I took one end between my teeth, plunged into the river, and swam to the other shore, towing him after me. In like manner we crossed the Colorado also. My knowledge of the country enabled me to avoid all public roads. Our journey led us through extensive prairies, and sometimes almost impenetrable forests.

On the eleventh day, having traveled about one hundred and fifty miles, we accidentally fell in with a soldier, who had just returned from the Battle of Jacinto. Here I was informed of the joyful news of the capture of General Santa Anna. Our means of subsistence, during the time of our flight, consisted of a few rations of bread that I had saved, and two small pigs found at deserted houses on our way. Continuing our journey, we found ourselves at Velasco, the present seat of government, about the middle of May. My health being much impaired, I obtained a furlough, with permission to visit my friends in the United States.

Charles B. Shain of Kentucky

[Extracted from C. B. Shain, writing at Louisville, KY on June 25, 1836, as published in the Louisville Journal *five days later. Shain fled with Thomas Kemp of the Mobile Greys, Daniel Murphy and perhaps 2 others, who did not leave any known written accounts.]*

On the 7th of February, I joined Captain Burr H. Duval's company and remained in it until the butchery of Colonel Fannin and his men. We went to work and fortified ourselves, pulling down and burning all the houses in the place. We went down the San Antonio River, on one occasion, after some suspicious fellows. We took nineteen of them prisoners and returned in a few days.

On the Friday previous to our final departure, our horsemen came in and gave the alarm that there were a good many persons in the woods opposite the fort. Col. Fannin ordered our company and the Red Rovers to cross the river and cover the retreat of the horsemen. We fired two or three rounds of artillery at them, but they soon made off. That night we intended starting after dark, but some of our horsemen came up from the river, and said that there was a picket guard of the Mexicans at each ford. Col. Fannin then ordered Col. Horton to take his horse company and cross over the river with one of our company behind each of them, and to watch until we could have the artillery and baggage carried over. We thought it a very singular order, but we obeyed.

The horsemen went forward, and, in a short time, one of them came galloping back, and told us that there were at least 200 horsemen in the act of crossing. In a few minutes we heard horses coming and were ordered to form and receive a charge. They came within fifty yards of us before we could see them on account of the darkness. Captain Duval hailed them, when we found them to be our own men that we had sent to see if there was any chance of crossing that night. We were very near shooting at them. One of our guns snapped; and if it had gone off, we should certainly have killed nearly every man, for we all had our triggers sprung and our rifles cocked. It was so dark that the Mexicans did not pursue us.

We then returned to the fort, and the next morning, at 11 o'clock we were across the river. We marched about six miles, when Col. Fannin ordered us to halt, and let the oxen rest. We stopped about one hour, or probably an hour and a half, in which time, had we proceeded, we could have reached the woods in safety. We had not marched more than two or three miles, when our flank guard came riding in, and said that the Mexican army was advancing on us. By this time they were in sight. We immediately halted and fired two or three cannon at them, but they were too far off to be injured. We then hitched our oxen and marched about a half mile farther when we saw a large body of Mexicans in our rear and on both sides, advancing on us very rapidly.

We halted, formed a hallow square and commenced again with our artillery, but they still advanced until within about 400 yards, when at about 3 o'clock, they commenced firing with their muskets but still continued to advance. They marched towards us slowly until they got within 150 yards. We then commenced with our rifles and muskets. As soon as we opened our fire they fell back about 200 yards, and we kept up regular fire until nearly sundown when they retreated.

It was then proposed by some of the officers, that we retreat to the woods, but some of the men objected on account of our wounded. We had about 20 or 25 men that would have been left. If our advance guard had not been cut off from us we could have carried our wounded and made our escape to the woods and water, where we could have whipped the enemy with all ease. They kept sounding their bugles every five minutes during the night, and we expected a charge every minute. A prisoner that had been taken at Mission Refugio said they were all night burying their dead. We had but six killed in the battle and about forty wounded. On the Mexican side, about 1,100 were killed or wounded; or at least that is the number Almonte says were missing next morning, but we supposed at that time that about 700 or 800 were killed and wounded.

That night the Mexicans were reinforced with artillery and about 500 men under the command of Colonel Bradburn, a Kentuckian, who had left Christian County, Kentucky, for stealing negroes.

Next morning about 8 o'clock, they fired three cannon and hoisted a white flag. We answered it immediately, and their commanding officer, Urrea, and two other officers, one a German, and the other a Mexican, came to meet us. Some of the propositions were: that they were to respect our private property, and that we were to be treated as prisoners of war until the expiration of eight days, when we were to be sent to the United States on parole of honor. Colonel Fannin then called all the officers together, stated the proposition to them, and a majority of then consented to it—in fact, I believe all of them consented with the exception of Captain Duval. He told them that they might do as they pleased, but he would never give his consent. The negotiations were concluded. We then stacked our arms and marched in double file back to the fort. We arrived there late in the evening.

They gave us nothing to eat that night and nothing till late next day. Then they gave us about as much to last twenty-four hours as we could eat at one meal. We were kept in an old church for two days and nights, after which we were kept in the fort until the next Sunday morning. On Saturday the 26th, six days after we were taken, Santa Anna arrived there, and we were the next morning taken out to be shot, but at the time we thought they were going to comply with the terms of the treaty.

They divided us into three divisions. The first division was led out on the Victoria road; the second, the division I belonged to, was taken out on the San Antonio road; as for the other, I do not know where they were taken, but I think that they were killed in the fort, as none of that division escaped. The division to which I belonged had proceeded as far as a brush fence, when a firing commenced in or near the fort. Our guards immediately ordered us to halt, but the rear had not halted before I heard somebody say, "Prepare!" The enemy all leveled their guns and fired. They were within three or four feet of us when they fired. They missed me and I ran to the river and swam over. While I was swimming they shot five times at me, at a distance of not more than fifty yards. John Duval, John Holliday, Daniel Murphy, Nat Hazen, and myself and several others swam the river together, but a good many were killed after they had reached the opposite shore. Murphy and myself met as soon as we crossed, but the others that crossed at the same time were killed on the other side. There were between 350 and 400 men killed, and only 16 escaped.

Murphy and myself had hid ourselves in the bushes, until the firing ceased, which lasted about two hours and a half. We then went about two miles to a small bunch of timber, and there we found John Williams in a tree. We concealed ourselves until after night, and then traveled northeast until it became so cloudy that we could not see our course. We stopped on the Coletta until morning. We then traveled all day, and nearly all the next night, though it was so cloudy we knew not which way we were traveling. On that day we fell in with two more of our companions, one of whom had belonged to the first, and the other to the second division. We went five days and a half without eating. On the sixth in the morning we found a small turtle. We immediately kindled a fire and cooked and ate it. It strengthened us very much, and shortly after that we came to the Guadalupe shore. We were pursued by six or eight Mexicans, but we saw them first, and ran to the river and swam it. We thought that we would hide in the bushes, but we found a very large panther in the bushes. The Mexicans in our rear, a large river before us, a panther in the bushes. Of the three dangers we took to the river and all got over safely and hid in the bushes on the other side until night.

We then crossed a large prairie and struck the Gonzales road, upon which we traveled until 12 or one o'clock. It was then raining, but I knew we must be somewhere very near Mr. Burnes' farm. We were in a thick wood, and stayed there until morning.

We started very early, and that morning, we got to Burnes' and found some of the best bacon I have ever tasted. We killed a hog that we found in a lot, cooked him, and parched enough corn to last about two or three days. That night we went five or six miles, and my companions had to stop for me, as I was very unwell. The next day we traveled for part of the day, and at night crossed the Gonzales, and struck the Texana road, which we followed until it crossed the San Felipe and Victoria road. We then took the latter, and followed it to the Labaca, when we went up the Labaca until we came to the settlements.

We got to Mr. Kent's on the 8th day after we left Goliad. There we met with four others. *[Survivor Samuel Brown, traveling with John Holliday, also mentions such a pit stop on the Lavaca at roughly the same time, but does not mention meeting any other survivors there.]* They were just eating supper when we got there. The people had left all their chickens which we killed and ate. On the following morning we thought we would go over to the next house and kill a hog, and cook him to carry with us. We stayed there four days for repose, it being a part of the country not much traveled. We finally started but lost our way, and had to come back about six miles, as we were determined to keep to the road. We traveled two days from Labaca before we got to the San Felipe and Gonzales road. The road looked very much like an army had, a few days before, passed along there. We had not traveled far before we saw two men on horseback coming towards us. We dodged out to one side of the road until they passed. We found them to be Mexican spies. We had a dog with us, which we had brought from the Labaca, and he came very near betraying us several times.

The next day we got to the Colorado. Before reaching the river, we met an old Dutchman, and brought him along with us. He was going directly towards the Mexican army. We told him of the butchery of all of Colonel Fannin's men, but he could not understand it. He said that he had heard there was to be war six months ago, but had never heard any more about it. He said all his neighbors had removed, and he was afraid of the Indians. When we got to the Colorado river, Murphy, Kemp, and myself, swam the river, and went to Lacy's four miles distance to kill a hog while the others made a raft to get the Dutchman across the river. We went on to Lacy's and got a very fine hog in a pen and killed him. While we were skinning him, we heard the Mexican drums.

The next morning, after traveling about four or five miles we came in sight of the Mexican camp in Tuskasett Prairie. We went into the woods and followed them around to the San Felipe road. We had not traveled far when we saw about 600 Mexican soldiers. We then got into the woods about a half mile off and staid there until night. That night we came upon the Mexicans encamped on the San [Bernard] river, and went up that river about five miles before we could get across on account of quicksand. We got in the road again and had come within four or five miles of San Felipe, when we came on a Mexican picket guard lying asleep in the road. We left the road without disturbing him, and went out in the prairie about two miles from where we had seen the picket. It being very cloudy, we could not travel so concluded to lie down in the prairie until the moon rose, but we went to sleep and did not wake till sunrise. Then we had about five miles to travel before we could get to the woods on the Brazos river. In doing this we went in sight of the Mexicans at San Felipe, who were sounding their bugles, beating their drums, and firing their cannon. We went through the timber on the Brazos to the river, which one of our company said he could not swim. We then went into the prairie along the edge of the woods which were too thick for us to pass through. We concluded to go to Washington and there cross the river and see if we could not hear any of the American army.

We went on to Mr. Cummings' house on Mill Creek, which was very high and we had to swim it. We had not been there more than an hour when we saw one of our spies. I called to him but he was a little backward at first, being some distance off at the time. He came up to us, and that night, rode back to

Campana, and sent us horses. That night we stayed in our picket guard camp, four miles from Cummings' and four miles from General Houston's camp.

The next morning we reached the Texan camp. As soon as I arrived, Colonel Benjamin F. Smith sent for me and gave me some clothes and told me that his negro boy would wait on me till I was well. We crossed the Brazos in four or five days after we got to the camp. We crossed at Groce's, and stayed there two or three days, and then took up our line of march for Harrisburg on Buffalo Bayou.

We arrived at Buffalo Bayou opposite Harrisburg on the 17th of April, and that evening our spies took three Mexican couriers with dispatches to Santa Anna. The Mexicans were in Harrisburg at the time we encamped on the other side. That night about eleven o'clock, we heard the Mexican drums beating, and the next morning, we heard that they had taken up their line of march towards Anahuac. We immediately crossed the bayou which took us all day, and the army marched nearly all night. The next day we came in sight of the Mexican army. Our army encamped on Buffalo Bayou near San Jacinto that night, we went back and crossed the bridge that crosses Sim's Bayou.

On the 20th the Mexicans commenced cannonading our camp. General Houston ordered Colonel Sherman to lead out the cavalry and take a piece of woods before the Mexicans could do so, but the Mexicans got there first and Sherman, not seeing them, marched slowly into the timber, when he was fired on by the Mexicans, but nobody was hurt. There was one horse killed.

On the next day, the men were all formed, and Gen. Houston told them that [any man who] could not stand the bayonet must stay behind. The left wing of our army was commanded by Col. Sherman, the center by Gen. Houston, and the right by Col. Burleson. The Mexican's army left was commanded by Col. Almonte, the center by Gen. Cos, and the right by Santa Anna. The Mexicans were in the wood, and had thrown up a breastwork, and we were in the prairie. Col. Sherman's command was led up under the brow of the hill, until they were within seventy or eighty yards. The fire was then opened, and they went in double-quick time over the enemy's breastwork. Our whole army was across the breastwork in fifteen or twenty minutes after the battle commenced. The Mexicans were then running in all directions, and our men either threw away their guns or used them as clubs, showing the enemy no quarter, at first. Our watchword was "Alamo and La Bahia." There must have been 200 killed in crossing the bay. The Mexicans had in all about 700 killed and between 700 and 800 taken prisoners. We had three killed, and thirteen wounded.

We stayed there twelve days after the battle and then going on board a steamboat, I went to Galveston Island.

Joseph Spohn, Interpreter

[Joseph Spohn's account, most often cited for his detailed account of Fannin's execution, was originally published in The New York Evening Star *in the summer of 1836, an extant copy of which this compiler is unable to locate. Extracts from the New York paper were reprinted on August 9, 1836 in a Pennsylvania paper. Taken here from O'Connor's* Presidio La Bahia.*]*

Mr. Joseph H. Spohn, one of the survivors of Fannin's command, has arrived in New York in the ship *Mexican* from Vera Cruz. He owes his escape to being able to speak the Spanish language, which made his services necessary as an interpreter to the savage Mexicans. Spohn was one of the Red Rover volunteers, and went to Goliad with Col. Grant, Col. Johnson and Major Morris, uniting there with Fannin's party, with the Georgia Battalion and Alabama Grays. He had furnished the *New York Evening Star* with many interesting particulars, given editorially in that paper, which begin with the battle and unfortunate surrender at Coleto, and come down to his escape from Vera Cruz. We make the following interesting extract, detailing the circumstances of the murder of his unfortunate friends:

On Palm Sunday, the 27th of March, the prisoners were formed into line, and Mr. Spohn, who was then sleeping in the church, (the hospital), about 6 o'clock in the morning, was called out and told to form into line. Being the last, he fell at the end. They were then marched out of the fort and ranged before the gate, when an officer stepped up and asked Spohn what he was doing there, and ordered him to go back to the hospital where he was wanted. While on his way, he was stopped by another officer, who told him to order the assistants to have the wounded of the Texians brought into the yard, such as could not walk were to be carried out. Being astonished at these preparations, he asked why, when the officer said, "Carts were coming to convey them to Copano, the nearest seaport." The orders of the officers were obeyed, and the wounded brought into the yard, and they were all full of the hope that they were to be shipped to the United States.

But their hopes were cruelly blasted when they heard a sudden continued roar of musketry on the outside of the fort, and observed the soldiers' wives leap upon the walls and look towards the spot where the report came from. The wounded were then conscious of what was passing, and one of them asked Spohn if he did not think that their time was come; and when they became convinced from the movements about the fort that they were to be shot, the greater part of them sat down calmly on their blankets, resolutely awaiting their miserable fate. Some turned pale, but not one displayed the least fear or quivering. Spohn, who was employed in helping them out, was accosted by a wounded Mexican soldier on whom he attended, and told to go and ask the commandant for his life, since they were all to be shot.

About this time Col. Fannin, who had a room in the church for his use, came out of the church for a particular purpose, when a Mexican captain of the battalion, called Tres Villas, with six soldiers, came up to Spohn, and told him to call Col. Fannin, at the same time pointing to a certain part of the yard, where he wished him to be taken. Spohn asked if he was going to shoot him, and he cooly replied, "Yes." When Spohn approached Fannin, the Colonel asked what was that firing, and when told the facts he made no observation, but appeared resolute and firm, no visible impression on Colonel Fannin, who firmly walked to the place pointed out by the Mexican captain, placing his arm upon the shoulder of Spohn for support, being wounded in the right thigh, from which he was very lame.

When Colonel Fannin reached the spot required, the northwest corner of the fort, Spohn was ordered to interpret the following sentence: "That for having come with an armed band to commit depredations and revolutionize Texas, the Mexican Government were about to chastise him." As soon as the sentence was interpreted to Fannin, he asked if he could not see the commandant. The officer said he could not, and asked why he wished it. Colonel Fannin then pulled forth a valuable gold watch, he said belonged to his wife, and he wished to present it to the commandant. The captain then said he could not see the commandant, but if he would give him the watch he would thank him—and he repeated in broken English, "Tank you—me tank you." Colonel Fannin told the captain he might have the watch if he would have him buried after he was shot, which the captain said should be done—"Con todas. Las formalidades necessarias"—at the same time smiling and bowing. Col. Fannin then handed him the watch, and pulled out of his right pocket a small bead purse containing doubloons, the clasp of which was bent. He gave this to the officer, at the same time saying that it had saved his life, as the ball that wounded him had lost part of its force by striking the clasp, which it bent and carried with it into the wound, and a part of a silk handkerchief which he had in his pocket, and which on drawing out drew forth with it the ball. Out of the left pocket of his overcoat, (being cold weather he had on one of India rubber) he took a piece of canvass containing a double handful of dollars, which he also gave to the officer.

Spohn was then ordered to bandage his eyes, and Col. Fannin handed him his pocket handkerchief. He proceeded to fold it, but being agitated he done it clumsily, so the officer snatched it from his hand and folded it himself, and told Col. Fannin to sit down on a chair which was near, and stepping behind him, bandaged his eyes, saying to Col. Fannin, in English, "Good, good?"—asking if his eyes were properly bound—to which Fannin replied, "Yes, yes." The captain then came in front and ordered his men to unfix their bayonets and approach Col. Fannin. Hearing them near him, Fannin told Spohn to tell them not to place their muskets so near as to scorch his face with the powder.

The officer standing behind them after seeing their muskets were brought within two feet of his body, drew forth his handkerchief as a signal, when they fired, and poor Fannin fell dead on his right side on the chair, and from thence rolled into a dry ditch, about three feet deep, close by the wall. They then led Spohn near the gate, from which another officer took him, and placed him in the room of Colonel Portilla, with a sentinel over him. He asked the officer if he was going to shoot him. The officer replied, "No, hijo," grinning maliciously at the same time.

In the room he found a Frenchman of the Copano Company, who told him the rest of his corps had, early in the morning, been placed in a garden, outside the fort, under guard. After Spohn had been there a short time, a soldier with his gun came to the door, telling him he was wanted at the gate. When he came to the gate he found Commandant Portilla surrounded by his officers, who, on seeing Spohn, begged Portilla to save him, but he said he could not, as his orders were positive. But the officers persisted, and he rather impatiently said, "Well, take him away."

At the same time he saw them lead young Ripley, second sergeant of the Mobile Greys, who was badly wounded in the left arm, to the place of execution. Spohn had been in the house but a little time when a young Mexican soldier with a bloody sword, entered the room and asked him what he was doing there, and would have run Spohn through had not the servants told him he was placed there by the officers. Dr. Field came in with a sergeant. The doctor told Spohn that all were shot, and they had roughly dragged Captain Brooks of West Point, who laid with his thigh broken, from a house outside the fort, and dispatched him brutally in the street.

In an hour more Spohn re-entered the fort, where he found the Mexican soldiers placing the bodies of the dead on a large wagon and carrying them off. Two or three days after, Spohn was taken by Captain Corono to the place outside the fort where his countrymen had been murdered and piled one upon the other, and partially burnt or roasted, presenting a most frightful, horrible, and disgusting spectacle. Here, he found that they had been divided into four parties before they were shot, as there were four piles, surrounded by torn pieces of bloody clothing, shoes, caps, pocket books, and papers. Among the rest was the bloody cap of

Fannin, which leads us to expect he was burnt or roasted with the others.

Five men were saved from the general massacre to attend upon the Mexican sick: Skerlock, Smith, Bills, Voss and Peter Griffin. The latter, who happened to be in the hospital at the time, was saved by a wounded Mexican soldier, who hid him beneath his blankets, Griffin having always attended him and dressed his wounds. Bills died afterwards, from a sickness of 24 hours.

We were conducted out to a peach and fig grove, in front of the church, and in sight of two of the three parties into which Fannin's men were divided; the third being out of view behind the church, near the riverbank. When the firing began, boy as I was, I was impressed by the varied expressions in the faces of our men, thus made unexpected witnesses of the awful tragedy. Surprise, horror, grief and revenge were depicted in the most vivid lines. At first all were startled; some became at once horror stricken; others wept in silent agony; still others laughed in their passion, swore, clinched their teeth, and looked like demons. Now, at the lapse of more than a quarter of a century, I can never think or talk of that dreadful scene with any degree of composure.

—S.H.B., Secretary to Captain Miller

Milton Irish Writes Home

[Milton Irish, a native of Maine, was a participant in Texas Revolution from beginning to end. A member of the Matamoros Expedition, he managed to evade capture and death at the Battle of Agua Dulce and make his way on to Goliad where he would become attached to the San Antonio Greys. He made his second escape from death at Goliad. Following the Revolution, Irish settled in San Augustine County. He married Emily Eaves in 1845 and served two terms as the San Augustine County Coroner. In 1852, he jumped on the Gold Rush bandwagon, making his way to California. He remained there, afterwards engaging in the logging trade, while his wife and children remained in Texas. He is believed to have died in San Mateo County, California in 1869, at the age of fifty-seven. He wrote to his father in Maine in early 1837 about his experiences in Texas. The letter was published the Lincoln County, Maine Patriot *on February 10, 1837. This extract picks up after his first escape. He is en route to Goliad.]*

...I therefore resolved to go to Labadea [La Bahia], as soon as possible, to procure assistance. On the road I fell in with the guard I had left in the thicket, on their way to Labadea. When I arrived within 9 miles of Labadea, we met a body of Americans on their way to the Mission, having received an express for assistance some hours before. I proceeded to Labadea where were about 300 men. The next day orders were received from Col. Fanning to abandon the place. An express was sent from the Mission for our men who were on the march to return to Labadea with all possible speed, as the former place had fallen into the hands of the enemy. A few days after having discovered a body of Mexicans encamped within 5 miles of us, we concluded to retreat to Victory [Victoria], a small town on the Warlope [Guadalupe], about 25 miles off; for which we took up our march on the day following the discovery.

After proceeding about 6 miles, we discovered a small party of the enemy on our left. We halted on a commanding eminence where we remained an hour or two. We proceeded a short distance from thence when we saw a large body of horse[men] on our right. We were immediately formed for battle and sent them a few rounds of canister when they made a manoeuvre to throw themselves in front of us, seeing which, our commander ordered us to march to the timber about a mile distant. The foe, seeing this, advanced upon us, both the cavalry and infantry, which had now come up.

A sharp engagement ensued, in which the Mexicans were repulsed with the loss of 200 men. Our loss was 7 killed and 63 wounded. This engagement lasted three or four hours in which they had greatly the advantage, being on the rising ground, while we were in a hollow, and they numbering also upwards of 2,000 men, while were not more than 300.

They had no artillery. Ours, consisting of 7 pieces, was silent some time before the close of the action, owing I suppose to the number of wounded of that department. My rifle having become useless, I repaired to a six pounder, and having procured several charges for her, with the aid of a cowardly Irishman and a brave Pole, I fired twice with good effect, on a body of horses advancing upon us.

Immediately after the repulse of the enemy, we were employed to dig a ditch, at which we continued most of the night. In the morning the Mexicans appeared with several pieces of artillery and large reinforcements. They hoisted a flag of truce, which was followed by a capitulation honorable for us to accept, throwing us into their hands. We were taken back to Labadea, closely guarded and allowed about 3/4 of a pound of fresh beef a day. Four days after, 60 prisoners were brought in who had been taken on their landing a Copono [Copano].

On the morning of April 2nd [March 27] we were paraded, for what purpose we knew not. We were then marched out of the Fort, divided into three parties and led, one company down the river, the 2d up the river, the third in which I found myself, was taken about 300 yards from the fort, into a pen of about a quarter of an acre surrounded by a high brush fence. We were marched to the opposite side of the pen, with a file of men on each side, when one file halted and we were led round, till our whole line was brought along the fence, and then halted. The manoeuvres first intimated to us what was to be our fate. Here Death for an instant stared me in the face. I inwardly cried, Lord have mercy on me! When a thought occurred to me: I had spent my life in wickedness, and it was now too late for hope. From that moment all fear left me; a desperate indignation took its place.

In that awful hour a few impassioned exclamations burst from the men such as "Lord have mercy! O Lord!" A young gentleman from Alabama said in a firm tone "Gentlemen, let us meet our doom like men."

A young man on my right beginning to cry, I said to him, "Let us break." He replied, "No it is useless to run we shall all be killed." At this moment I heard the order for our executioners to make ready, and saw their guns presented not ten feet distant; yet such was the state of my feelings that I remained unmoved. I heard the order to fire, the roar of musketry followed; still I remained unhurt.

But what a scene presented itself to my view! Those death shrieks still ring in my ears. Glancing my eyes hastily around, I discovered on my left, about a dozen men who had made their way over the fence. I sprung for the fence as one springing for life. I fell prostrate but, recovering, was soon on the other side. A widespread prairie was now before me, and the only shelter which offered was to the right or left where the other divisions had been marched. I chose the right. After running about 100 yards I ventured to look over my shoulder. A tall officer with a drawn sword was in pursuit of me. I then threw off my hat and coat. Again looking behind, I saw that my pursuer had stopped.

As I passed near the main part of the town, I saw a horseman start after me. Perceiving this I shifted my course more to the right, for the heads of the valleys lay between the town and the division of men I had to pass. On this the horseman wheeled and took after some men running in the prairie. The division before me was now fired on; and seeing several men run nearly for me, I shifted my course, and soon succeeded in reaching the bushes. I bent my course for the river, which I crossed, and proceeded some distance on the bank, when I again swam the river, as I had done five times before, and concealed myself where I remained till dark, when I ventured once more upon the prairie.

By the reason of the great exertions I had made, being unable to proceed, I was obliged to lie down beneath the shelter of a single bush, where I soon fell asleep. How long I slept I know not, as I found myself standing on my feet, awakened, as I thought by the report of a musket; which was frequently the case afterwards. I bent my course for the American settlements, and after 10 days reached the Colorado having seen many of the enemy at a distance, at different times. Here I found the campfires of the Mexicans still burning. I found also a calf's head roasted by them the preceding evening, which was the first cooked food I had taken since my escape, having subsisted on raw vegetables.

In 2 days I reached the River Brazos which was swollen to such a degree that I dare not attempt to cross it. I here heard the enemy crossing and often saw their scouts, so that I deemed it not safe to go in quest of food. I therefore returned to the Colorado, where I often saw parties of both Mexicans and Indians. My concealment lasted more than six weeks during which time I procured plenty of provisions from neighboring farms. On the eighteenth of May I discovered that the Americans had returned to their farms, whom I hastened to meet, being the first persons to whom I had spoken since my escape. From them I learned that Santa Anna was a prisoner; also that the Texian army would be there in 2 days.

On the arrival of the army, I again joined it and went to Labadea, where I witnessed the funeral services of my former companions, whose remains after the massacre had been partly burned. This was by far the most trying scene through which I ever passed. I continued in the army till June 2d, when I procured my discharge, and reached San Augustine the 25th of the same month.

"The prisoners of Goliad were condemned by law, by a universal law —that of personal defense— enjoyed by all nations and all individuals."

—Antonio Lopez de Santa Anna in *Manifesto*, 1837

Benjamin H. Holland, San Antonio Greys Artillery Captain

[Captain Benjamin Holland's account of the action at Goliad is taken from Pease's Geographic and Historical View of Texas, *1837.]*

...We accordingly met in council to devise means and measures; it was accordingly decided that we should send a flag of truce to the enemy, and if possible, obtain a treaty, if upon fair and honorable terms. Accordingly Capt. F. J. Delanque, the bearer of the express from General Houston, Capt. B. H. Holland of the artillery, and an ensign, were despatched with a flag of truce. The flags met midway between the two armies, and it was decided that the two commanders should meet to decide the matter—in pursuance of which Col. Fannin was conveyed out and met Gen. Urea, Governor of Durango, commander of the Mexican forces, and the following treaty was concluded upon, and solemnly ratified; a copy of it in Spanish was retained by Gen. Urrea, and one in English by Col. Fannin:

Seeing the Texian army entirely overpowered by a far superior force, and to avoid the effusion of blood, we surrender ourselves prisoners of war, under the following terms:

Art. 1st. That we should be received and treated as prisoners of war, according to the usages of civilized nations.

Art. 2nd. That the officers should be paroled immediately upon their arrival at La Bahia, and the other prisoners should be sent to Copano, within eight days, there to await shipping to convey them to the United States, so soon as it was practicable to procure it, no more to take up arms against Mexico until exchanged.

Art. 3d. That all private property should be respected, and officers' swords should be returned on parole or release.

Art. 4th. That our men should receive every comfort, and be fed as well as their own men.

[Signed]

Gen. Urrea
Col. Morales
Col. Holzinger

on the part of the enemy, and on our part signed by Colonel Fannin and Major Wallace.

The officers were then called upon to deliver up their side arms, which were boxed up, with their names placed by a ticket upon each, and a label upon the box, stating they should soon have the honor of returning them, and it was their principle to meet us now as friends, not as enemies.

Col. Fannin and the men were that afternoon marched back to La Bahia; the wounded, together with the captain of each company, and our surgeons, were left on the field to dress the wounded, which was completed on the 21st, when we were all conveyed back to the fort, where we found the men in a most miserable state. They were brutally treated—they were allowed but very little water to drink, in consequence of its having to be brought from the river, and but a small piece of meat, without salt, bread or vegetables. On the 23d, Major Miller and ninety men were brought into the fort as prisoners; they had just landed at Copano from the United States.

On the 25th the Georgia Battalion was also brought in. It had been surprised and captured between Victoria and Dimmit's Point, and marched back and confined with us. Here we were now nearly 500 strong, guarded by 1,000 Mexicans, without being allowed the slightest liberty in any respect.

The Mexicans had always said that Santa Anna would be at La Bahia, on the 27th, to release us. Accordingly, on that day, we were ordered to form all the prisoners; we were told that we were going to bring wood and water, and that Santa Anna would be there that day. We were ordered to march all the officers at the head of the file, except Col. Fannin, who lay wounded in the hospital. As we marched out of the sally port we saw hollow squares formed ready to receive us. We were ordered to file left, and marched into a hollow square of double-filed cavalry, on foot, armed with carbines, commonly called scopets, and broad swords.

This square was filled and closed, and the head of the remaining files wheeled off into another square, and so on, until all were strongly guarded in squares: the company of which the writer of this was one, was ordered to forward, and no more was seen of our unfortunate comrades. We marched out on the Bexar Road, near the burying ground, and as we were ordered to halt, we heard our companions shrieking in the most agonizing tones, "Oh, God! Oh, God! Spare us!" and nearly simultaneously a report of musketry. It was then we knew what was to be our fate. The writer of this then observed to Major Wallace, who was his file leader, that it would be best to make a desperate rush—he said no, we were 100 strongly guarded; he then appealed to several others, but none would follow. He then sprung and struck the soldier on his right a severe blow with his fist. They being at open files, the soldier at the outer file attempted to shoot him, but being too close, was unable. The soldier then turned his gun and struck the writer a severe blow upon the left hand. I then seized hold of the gun and wrenched it from his hand, and instantly started and ran towards the river. A platoon of men (I have been since informed, by two others who made their escape by falling when fired upon among the dead bodies of their comrades,) wheeled and fired upon me, but all missed.

I then had a chain of sentinels to pass at about 300 yards distance. They were about thirty yards apart, three of them closed to intercept my retreat, the central one raised his gun to fire—I still ran towards him in a serpentine manner in order to prevent his taking aim—I suddenly stopped—dropped my piece, fired, and shot the soldier through the head and he fell instantly dead. I ran over his dead body, the other two firing at me but missing, and immediately ran and leaped into the river, and while swimming across was shot at by three horsemen, but reached the opposite banks in safety; and after wandering six days without food in the wilderness, succeeded on the tenth of April in joining General Houston's army, after having been retaken by the enemy once, but succeeded in making my escape in company with a wounded man who had got off from La Bahia, by falling among the dead as before stated. I am happy to state, that six more succeeded in saving their lives and regaining their liberty by the same stratagem. The number of the enemy according to their own account, killed at the battle of Coleto, varies from nine to eleven hundred.

Samuel Brown - Survivor Recaptured

[Samuel T. Brown was a nephew of Col. William Ward of the Georgia Battalion. He penned this letter to Col. Ward's brother, Thomas, in 1837. It was published in the Alabama newspaper, Voice of Sumpter, *November 28, 1839, and reprinted in* The Texas Almanac for 1860.*]*

Livingston, Alabama
November 1, 1837

Dear Sir: Having been among the first who volunteered from Georgia in the service of Texas, under the command of your brother, the late Col. William Ward, whose name is destined to occupy a place in history, I have thought that a communication of my adventures in a form you might preserve, would not be unacceptable or improper. All I have in view is to give the facts which came within my own observation and knowledge; and if they can be deemed of interest as occurring to one of my years, (twenty at the present time,) I shall feel perfectly satisfied in having related them.

About the twentieth November, 1835, I left Macon in the stage for Columbus, where I joined Capt. Ward's company, which had rendezvoused at that place, from whence we marched to Montgomery, Alabama, and took passage for Mobile on the steamer *Ben Franklin*. Remaining in Mobile five or six days, near which a public dinner was given us, we embarked on the steamer *Convoy* for New Orleans, where we halted about a week, and received some addition to our number, making the company about a hundred and fifty strong. Here Capt. Ward laid in supplies for his men, and chartered the schooner *Pennsylvania* to take them to Velasco, where we arrived on the twentieth of December, 1835, and found Capt. Wadsworth's company, fifty strong; and the two companies were organized into a battalion, of which Capt. Ward was elected major, called the Georgia Battalion. Capt. Ward's original company was divided into two equal parts, as near as practicable, the command of one of which was given to Capt. Uriah J. Bullock, of Macon, and that of the other to Capt. James C. Wynne, of Gwinette County.

Major Ward lost no time in reporting in person his battalion to Gov. Smith at San Felipe de Austin. Our troops encamped about two miles from Velasco, on the Brazos river, where they subsisted on the two months' provisions laid in at New Orleans. After a week's absence to the seat of government, Major Ward returned with commissions for the several officers. We remained in the camp near Velasco, until first February, 1836, when the battalion was ordered by the then acting Gov. Robinson, to repair to Goliad on the San Antonio River, and it was forthwith transported by the schooner *Columbus*, U. S. vessel, to Copano, on Aransas Bay, after five days' passage. There we were furnished with supplies by the government and four pieces of artillery, two six and two four-pounders. From Copano to Goliad the distance is forty-five miles, and about halfway the battalion halted at the Mission, where we were joined by Capt. Ticknor's company of Montgomery, Alabama, making our ranks about two hundred and fifty strong. From there we marched to Goliad, took possession, and repaired the Fort, and were joined by the Lafayette Battalion, made up from northern Alabama, Tennessee, and Kentucky.

Previous to this, the lamented Col. Fannin had not taken any part in service, but was actively engaged in collecting and diffusing information highly useful to the cause of Texas. At Goliad the two battalions were formed into a regiment, between five and six hundred strong, of which Fannin was elected Colonel and Ward, Lieutenant Colonel; Dr. Mitchell of Columbus commanded the battalion, in the place of Major Ward, promoted.

For some purpose, Capt. King, of the Lafayette Battalion had been dispatched by Col. Fannin to occupy the Mission, about twenty-two miles off, who found himself annoyed in his new position by a party of Mexican cavalry, and sent an express to Goliad for a reinforcement. Lt. Col. Ward, with one hundred and twenty men, of which I was one of the number, was directed by Col. Fannin to support Captain King at the Mission. This was on the twelfth March, and the next day Lt. Col. Ward's command reached the Mission, at which a large Catholic church built of stone made a very good fort, in which we took protection. The Mexican cavalry that reconnoitered the Mission and tried to attack it, was estimated at two hundred, and on the night of the nineteenth, a party of fourteen men under Capt. Ticknor, surprised their camp, a mile from the Mission, killing eight of them and putting the rest to flight. Among the slain was recognized a Mexican lieutenant who had been with Col. Fannin at Goliad, pretending to have joined the Texans with eighteen men. On the morning of the sixteenth, Lt. Col. Ward and Capt. King differed as to who should command at the Mission, the latter claiming it by being there first. A large majority of the troops declared they would serve under Lt. Col. Ward only, which induced Capt. King with his original company of twenty-eight men to withdraw, and was followed by eighteen of Lt. Col. Ward's command, who had been detailed from Capt. Bradford's company at Goliad, leaving Col. Ward one hundred and seven men. About ten o'clock in the morning, a party of fifteen with myself, was sent to a river about two hundred yards off, with oxen and cart, to bring two barrels of water into the fort. We had just filled the vessels and were leaving the river when we were fired upon from an open prairie on the other side, by General Urrea's army of eleven hundred men, about half a mile distant. We made all possible speed for the fort, holding on to the water, except about half a barrel, which was let out by one of the bullets piercing the head. The enemy kept firing as they crossed the river, and marched within fifty paces of the church, when Col. Ward ordered his men to fire, which drove the Mexicans back and left the ground pretty well spotted with their dead and wounded. They made four regular charges, both cavalry and infantry, about half of each, and were as often repulsed with great slaughter.

At four o'clock in the afternoon they retreated, leaving between four and five hundred of their dead upon the field. Col. Ward had only three of his men wounded, one of them an Irishman who resided at the Mission. When the attack was made in the morning, Col. Ward sent an express (James Humphrey, of Columbus, Ga.) to Col. Fannin at Goliad; and orders were received at ten o'clock at night, to abandon the church, take a northeast course for Victoria, on the Guadalupe, twenty-five miles beyond Goliad, where Col. Fannin would join him. About twelve o'clock at night we left the fort silently, formed five deep, marched without a guide in the open prairie, and were only eight miles from the Mission at daylight. For two days we had nothing to eat, and on the third we killed some cattle near the San Antonio, which revived us a good deal. On the twenty-first of March we reached Victoria, and had advanced within one hundred yards of the town, expecting to find Col. Fannin and his men there, when to our utter dismay it was in possession of the enemy, who fired upon and caused us to retreat to the swamp. Col. Fannin had set out to meet us in due time, but his whole command was taken prisoners by a large force within six miles of Goliad, and carried back to the fort. We had expended all our ammunition at the battle of the Mission, and very few of our men had a single cartridge! In this dilemma we marched a night for Dimmit's Point on the La Bacca river, near Matagorda Bay, where supplies were landed for the Texan troops.

Next day, twenty-second March, we halted to rest and conceal ourselves within two miles of our destination, sent two men to the Point to see who was in possession and await their return. The remnant of the Mexican army that attacked the Mission, and was hovering over this quarter under Gen Urrea, took the two men prisoners and surrounded us. The two men came within speaking distance of us, stated our situation and the power of the enemy, and desired Col. Ward to see Gen. Urrea upon the terms of surrender: upon which Col. Ward, Major Mitchell, and Capt. Ticknor, had an interview with Gen. Urrea and returned, making known to us the offer of the enemy, if we surrendered prisoners of war, that we should be marched to Copano without delay, and from thence to New Orleans, or detained as prisoners of war and be exchanged. Col. Ward addressed his men and said he was opposed to surrendering, that it was the same enemy we had beaten at the Mission, only much reduced in numbers, and that he thought our chance of escape equally practicable as it was then. He proposed that the attack on us might be evaded until night, when he might possibly pass the enemy's lines and get out of danger. At all events, he thought it best to resist every inch, as many of us as could save ourselves, and if we surrendered, he had doubts of the faith and humanity of the Mexicans; that he feared we should all be butchered. The vote of the company was taken, and a large majority were in favor of surrendering upon the terms proposed; Col. Ward informed them that their wishes should govern, but if they were destroyed, no blame could rest on him.

The same officers as before, to wit: Col. Ward, Major Mitchell, and Capt. Ticknor, again saw Gen. Urrea, and I understood a paper was signed by the Mexican General, to dispose of us as above stated, on condition that we should never serve Texas any more; one copy in Spanish and another in English. Then came the hour for us to see all our hopes entirely blasted. We marched out in order and grounded arms, cartouche-boxes, and weapons of every kind. Our guns were fired off, the flints taken out, and returned to us to carry. When we left the Mission, on the night of the 14th of March, we had about a hundred men; at the time of the surrender we had only eighty-five, the others having left us on the route from the Mission to Victoria—a most fortunate thing for them. We were put under a strong guard, and the next morning, 23d March, proceeded to Victoria, where we were engaged the next day in bringing the baggage of the Mexican army across the Guadalupe, about four hundred yards from the town, and hauling it up. On the morning of the 22d, we were marched toward Goliad, where we arrived next day late in the evening. There we found Col. Fannin and his regiment prisoners in the fort. All the

Texian troops then in the fort as prisoners, belonging to Fannin's command, after we were brought in, amounted to four hundred and eighty men. Early on the morning of the 27th, we were all marched into line and counted, and divided into four equal parts of one hundred and twenty each. The nearest to the door of the fort marched out first, and were received by a strong guard and placed in double file, going we knew not whither nor for what purpose. I was in this division, in the right-hand file, and about half a mile from the fort we were ordered to halt; the guard on the right then passed to the left, and instantly fired upon the prisoners, nearly all of whom fell, and the few survivors tried to escape by flight in the prairie and concealing in the weeds. The firing continued, and about the same time I heard other firing towards the fort and the cries of distress.

At the time our division of prisoners was shot, Drury H. Minor of Houston County, Ga., immediately on my left, was killed; and just before me, next in file, Thomas S. Freeman, of Macon, was killed. As I ran off, several poor fellows, who had been wounded, tried to hide in the clump of weeds and grass, but were pursued, and I presume killed. Soon after I made my escape, I was joined by John Duval and John Holliday, of the Kentucky volunteers, both of whom were with me at the massacre, but not until I had swam across the San Antonio, about half a mile from the butchery.

For five days we had nothing to eat except wild onions, which abound in the country; when reaching the Guadalupe found a nest of young pigs, and these lasted us several days. In the course of a few days, wandering at random in the open country, often wide off of our supposed direction, we saw fresh signs of cavalry, and withdrew to the swamp, but had been perceived going there, and were taken by two Mexicans armed with guns and swords; that is, Duval and myself were captured; Holliday lay close and was not discovered. One of the men seized me and held on; Duval was placed between them to follow on. He sprang off, and one man threw down his gun and ran after him in vain. Duval made his escape, and I have not seen him since. I was taken to their camp close by, when they saddled their horses in a hurry and rode off without me. From their actions I judged they were of opinion a party of Texians was near, and so made off. I then went to the swamp where I was taken, and found Holliday in his old position. Next day we came to a deserted house on the La Bacca River, apparently that of an American settler, where we found plenty of provisions, such as meat, corn, lard, chickens, and eggs, upon which we feasted there two days, camping at night a little way off. Taking with us a good stock of provisions, we traveled quite refreshed, and in four days reached the Colorado. From almost constant rain and exposure, I had lost the use of my right arm and shoulder, and could not swim the river. Holliday swam across with the provisions, and promised to return and help me; but he was so weak and exhausted from the cold and rapid current, that he was not able to do so. Thus we parted, and I never saw him afterwards.

I went up the river, and next day found a canoe in which I crossed, and then wandered till I got sight of the Brazos, on the 20th April, where I was taken by a party of twenty Mexican cavalry, who carried me to the main army at Ford Bend, under Gen. Sesma, and put me under guard with other prisoners they had picked up. I recollect the names of but three of them, and they had resided several years in Texas: Johnson, from New York, Leach, an Englishman, and Simpson. Fort Bend was about thirty miles from San Jacinto, where the battle was fought the next day, 21st April. The night after the battle a Mexican officer, who escaped from San Jacinto, brought the news into camp, and the army instantly retreated. When I was brought to the camp, I pulled off my boots to dry and relieve my swollen feet; my boots were stolen, and I had to march barefoot through the mud and water, nearly knee-deep all over the prairies, the rain falling in torrents pretty much all the time. The army returned to Victoria, where I saw four of the Macon company, who had been detained there after the surrender, on account of their being mechanics:

William Wilkinson, John C. P. Kinnymore, Barnwell, and Callahan.

I was then taken to Goliad, where I remained five days, and saw the places where the four divisions of prisoners had been butchered; some of the carcases remained, many burnt, and others mangled; all so disfigured that I could recognize no particular person. A company of eighty-two men, from Tennessee, under Capt. Miller, of Texas, who had been taken prisoners the moment they landed at Copano, and whom we left in the fort at Goliad at the massacre, still remained there on my return. One of its members, Mr. Coy, told me the particulars of Ward and Fannin's death, as he said he was an eyewitness. After all the men had been shot, the time of the officers came. Col. Ward was ordered to kneel, which he refused to do; he was told if he would kneel his life might be spared. He replied, they had killed his men in cold blood, and that he had no desire to live; death would be welcome. He was then shot dead. Col. Fannin made an address to the Mexican officer in command, through an interpreter; handed him his gold watch, to be sent to Col. Fannin's wife, also a purse to the officer to have him decently buried. He sat on a chair, tied a handkerchief over his eyes, and requested that be might not be shot in the head, and that the marksmen should stand far enough off for the powder not to burn him. He was shot in the head and expired.

Leaving Goliad in the month of May, with a dozen other Texian prisoners, under a guard of cavalry attached to the main army, then about three thousand strong, we marched to San Patricio on the Nueces river, where Cols. Teale and Carnes, of the Texian service, came under a flag of truce, and obtained passports from General Felisola to go to Matamoros, where Col. Teale informed me I should be discharged. I was kept with the main army, until Gen. Felisola received orders from Mexico to hasten there. He took with him a bodyguard through the Indian country, about fifty cavalry, who had charge of me ever since leaving Goliad, and they still held on to me. General Felisola left his guard at Saltillo, and took the stage to the City of Mexico, where the cavalry arrived with me, their only prisoner, in August, 1836. I was then confined in the *quartede*, or barracks, until the first of February, 1837, and about that time Gen. Felisola expected to leave the city to take command of the army at Matamoros. His interpreter, an Italian named Quarri, often visited the barracks, and treated me with great humanity. He said he would get my release, and took me to Gen. Felisola's house to accompany him to Matamoros. From some delay he did not start until the 28th of March, during which time I was a member of the family and treated with perfect kindness, under orders, however, (for my own safety, it was said,) not to leave the guard alone.

I may be allowed to say a few words about the City of Mexico and the manner of my detention. I was put in the barracks among a number of Mexican prisoners, who were confined for various offenses; and from the time I entered, in August, 1836, until I went to Gen. Felisola's house, in February, I had no other food than boiled beef.

The water in the barracks was fresh and pure, brought there by an aqueduct which supplies the whole city twelve miles from the mountains. The city itself is quite pleasant, clean, and the buildings durable, if not elegant. What I viewed as a great blemish to the houses, (which were nearly all of stone and rock,) were the images of saints and idols carved in endless variety.

On the 25th March last I left the City of Mexico in company with Gen. F., his staff, and a small guard, and arrived at Matamoros the first of June, a distance of nine hundred miles from one place to the other. Gen. F., it was said, declined the invasion of Texas with his army, on hearing of the death of Gen. Montezuma at San Luis, and sent a large portion of it to quell the insurgents. On the 17th June, 1837, Gen. F. gave me a passport, and on the 1st of July, I embarked for New Orleans on the schooner *Comanche*, Capt. Briddle, where I arrived in due time.

This unpretending narrative is at your service, and you have my permission to make what use of it you think proper.

I am, very respectfully, your obedient servant, —S. T. Brown

"Ate the matter from my wounds and tried to suck subsistence from green flies and lice of which I had a heavy stock."

—Survivor Isaac D. Hamilton in 1858

Captain Jack Shackelford

[In Henry Stuart Foote's 1841 history, Texas and The Texans, *he recounts, in a footnote, being closely acquainted with Captain Shackelford and tells of witnessing Shackelford's heartrending return home following the Goliad Massacre. Both gentlemen were calling Alabama home prior to succumbing to the allure of Texas. Foote also includes the Captain's narrative of the events at Goliad in his history. Both Foote's comments and the narrative are included here.]*

Foote's Comments on Shackelford's Homecoming

Captain Shackelford has been long most intimately known to me. He has enjoyed, for many years past, a standing and popularity in the state of Alabama, of which any of his contemporaries either there or elsewhere might feel justly proud. I would dwell in detail upon the many high and splendid qualities which adorn and beautify the character of Captain Shackelford both as a private and public man, were I not apprehensive of wounding that singular delicacy of mind with which I know him to be imbued, and which would render even deserved commendation, from one standing to him in the relation in which I occupy, more annoying than agreeable.

I feel bound to mention one fact, though. Nearly five years since, I chanced to be sojourning, for a few days, upon that romantic mountain known as the classic site of the promising College of La Grange; and one morning, I was standing upon a lofty eminence, looking down with delight upon the beautiful valley of the Tennessee, which expands here to the east and west as far as the eye can reach; when my ears were surprised with the unwonted sound of artillery, which seemed to be proceeding rapidly along the line of the railway that connects the towns of Tuscumbia and Courtland, like a "young volcano" in motion.

I inquired of those who were near me into the cause of what had awakened my surprise. What was my delight to learn that my old and dearly-loved friend, Dr. Shackelford, the renowned though unfortunate captain of the valiant Red Rovers, the companion in arms of the thrice-glorious Fannin, had just returned. His fellow citizens of the river bank were escorting him rejoicingly to his own home with military honours!

I afterwards heard that on his arrival in Courtland, he found a vast multitude convened to receive him, as it were, *from the dead.* Among these were the *Fathers* and *Brothers, Mothers* and *Sisters,* of those noble young heroes who had been lately committed to his charge, and whose bones he had been fated to see interred in the distant wilderness. All had come out now to congratulate him on his wondrous escape. But when he gazed upon the crowd, and remembered the *past,* his sensibilities were overpowered and he burst into tears, and all around him wept in unison. Never can that delicious yet doleful season of "mirth in funeral, and dirge in marriage," pass away from the recollection of those who were present.

Some Few Notes Upon a Part of the Texan War
by Jack Shackelford

Having lost all my papers and memoranda, I am unable to give precise dates, or to go minutely into detail. I promise, however, to give the truth in substance.

Some time in the early part of March, 1836, Col. Fannin had under his command at Goliad upwards of 400 men, consisting of Ward's Battalion from Georgia, and the following companies under the command of Major Wallace of the Texan army, who had recently been elected Major of the 2d Battalion, composed of the following companies, vis: New Orleans Greys [San Antonio Greys], Capt. Pettis; Mustangs, Capt. Duval of Kentucky; Mobile Greys, Capt. McManeman; Huntsville Volunteers, Capt. Bradford; [Auxiliary] Volunteers, Capt. King; and Red Rovers from Alabama, Capt. Shackelford. In addition to these, there was a regular company of Artillery, Capt. Westover; Hurst, Screnichi and Cornika (Polanders,) and Moore, Capt. of Guns. The companies were all small, excepting the Red Rovers, which numbered nearly seventy.

About the 12th of March, Capt. King's company was sent to the Mission of Refugio for the purpose of bringing off some families that were in a state of alarm. At the Mission, King encountered a large force of the enemy. Having taken protection in the church, he despatched a message to Fannin and, with his little band of 28 men, maintained himself against a large party of the enemy. About midnight on the 14th, King's express reached Goliad and Col. Fannin immediately despatched Col. Ward's battalion to his relief. This was the beginning of our trouble; and the only act for which I ever blamed Fannin. Those families should have left the Mission before they did, and Fannin should not have divided his forces; but that he was actuated by the best feelings, none can deny.

Ward reached the Mission on the following evening and cut his way through a large force, against which King had been gallantly contending all day. The next day the enemy withdrew some distance across a small stream, and were pursued by Ward and King, who unfortunately separated. This event led to the capture of King and his company, who were, as I have been informed by one present, marched a short distance and massacred in a cold-blooded manner; King meeting his fate with the intrepidity of a solider.

Ward returned to the church and, after having expended the greater part of his ammunition, retreated silently. Under cover of night, he made his way to the east in the direction of the Guadalupe. This manoeuvre eluded the vigilance of the Mexicans, as they had laid an ambuscade for him in the direction of Goliad. He reached Victoria on the 21st, after great suffering and being four or five days without anything to eat. At this place he expected to find the Texan army, and was not apprised of his mistake until surrounded by a large force of Mexicans under General

Urrea. One of his men, who made his escape, has assured me that Urrea *capitulated* with Ward, and *pledged* himself to afford him every *guarantee* according to the usages of civilized nations; and that even then, Ward was unwilling to capitulate until a majority of his officers consented to do so.

On the morning after Ward left for Goliad for the Mission, to relieve King, Col. Fannin received Gen. Houston's order to evacuate Goliad and fall back on Victoria. He took immediate steps in making preparation to *obey this order*, by dismounting several guns and burying them, sending out one or two parties of men accompanied by officers to procure teams and carts, and making other arrangements for an immediate retreat. An express was likewise forthwith sent to Ward, commanding him to return with as little delay as possible, and stating to him the nature of Gen. Houston's order. This express was followed by another, and yet another, who were all taken prisoners by the enemy; and it was not until the evening of the 18th that we received intelligence from Ward, and that not of a satisfactory character. I have mentioned this circumstance, if possible, to dissipate an unworthy prejudice which has been created in the minds of many that Fannin wished to forestall Houston in the command of the army, and therefore *disobeyed his orders*. I have said that he committed an error in separating his forces. Had he not done this, we should have been prepared to fall back on Victoria, as ordered, with a force sufficient to content with every Mexican we might have encountered. Fannin's great anxiety alone, for the fate of Ward and King and their little band, delayed our march. This delay, I feel assured, was not the result of any wish to *disobey orders*.

On the 16th of March, Col. Albert C. Horton of Matagorda, with twenty-seven men under his command, arrived at Goliad, bringing with them some oxen to enable us to take off our stores and munitions. A fourth messenger was despatched to Col. Ward, urging his immediate return, while we were busied in making preparations for a retreat. On the 17th, Horton was ordered to examine the country towards San Antonio and to keep scouts out in every direction. On his return, Horton reported a large force, a few miles from the fort, moving on slowly and in good order. We immediately dug up our cannon, which had been buried, and re-mounted them, expecting an attack that night or early the next morning. During the night, the guard was doubled and every arrangement made by the commanding officer to prevent surprise.

On the 18th, the enemy was still roving about the neighbourhood of the fort, and during the day, a large reconnoitering party showed themselves on the opposite side of the river, in the vicinity of the old Mission. Horton was immediately sent over with his company and a few others who could procure horses. I posted myself on the commanding bastion of the fort where I had a full view of the encounter that ensued. Horton behaved in a very gallant manner, and made a furious charge upon the enemy, drove them into the timber, and after encountering a very large force of infantry, fell back and formed his company in good order, immediately in front of the Mission. In this rencontre, young Fenner, of my company, shot a spy glass from the hands of an officer.

When I saw Horton in the midst of such peril, contending against such fearful odds, I obtained Col. Fannin's permission to go with my company to his relief. Such was the enthusiasm of the men that they waded the river up to their armpits, although by taking a little more time, we could have availed ourselves of the benefit of a flat which was at the ford. So soon as we reached the Mission and were about to flank the enemy, they made a precipitate retreat into the woods, although they outnumbered us ten to one. This was, no doubt, in part the result of a cannonading from the fort, which unfortunately commenced about this time. I say *unfortunately*, for we had every advantage of position and, could we have met even that force on such terms, I should not have feared the result. The cannonading from the fort was done at the insistence of the officer commanding the guns.

On the morning of the 19th, we commenced the retreat very early, the Red Rovers leading the van and Duval's company covering the rear. The lower road had been well examined by Horton's videttes, who reported all clear. At the lower ford of the San Antonio, much time was consumed in consequence of the inability of the team to draw our cannon up the bank. I waded into the river myself, with several of my company, assisting the artillerists by putting our shoulders to the wheels and forcing the guns forward. We then moved on briskly and in good order, Horton's scouts examining the country in front and rear. We had advanced about six miles when our scouts came in with a report that the route was still clear.

As our teams had become somewhat weary and very much in want of food, from having been kept in the fort for the last twenty-four hours, Col. Fannin determined to halt and graze them, and that we also might have time to take a little refreshment. I remonstrated warmly against this measure, and urged the necessity of first reaching the Coleto, then about five miles distant. In this matter I was overruled, and from the ardent manner in which I urged the necessity of getting under the protection of timber, I found the smiles of many, indicating a belief that at least I thought it prudent to take care of number one.

Here let me state one thing, lest I be misunderstood: Col. Fannin and many others could not be made to believe that the Mexicans would dare follow us. He had too much contempt for their prowess and too much confidence in the ability of his own little force. That he was deficient in that caution which a prudent officer should always evince, must be admitted; but that he was a brave, gallant an intrepid officer, none who knew him can doubt.

We halted near an hour, and then took up our march. Horton's Company was sent in advance to examine the pass on the Coleto. We had advanced about four miles, when a large force of cavalry were seen emerging from the timber, about two miles distant and to the west of us. About one half of this force (350 men) were detached and thrown in front of our right flank, with the intention of cutting us off from a skirt of timber, about one and a half miles in front. Our artillery was ordered to open upon them and cover our rear. Several cannon were fired at them, but without effect.

About this time, we discovered a large force of infantry emerging from the same skirt of woodland at which their cavalry had first been seen. Our guns were then ordered to be limbered; and we had purposed to reach the timber in front, but the enemy approached so rapidly that Col. Fannin determined to make an immediate disposition for battle. The prairie here was nearly in the form of a circle. In front was the timber of the Coleto, about a mile distant; in the rear was another strip of timber, about six miles distant; whilst on our right and left, equidistant, four or five miles from us, there were, likewise, bodies of timber. The order of battle was that of a hollow square. But, unfortunately for us, in endeavouring to reach a commanding eminence in the prairie, our ammunition cart broke down, and we were compelled to take our position in a valley, six or seven feet below the mean base, of about one-fourth of a mile in area.

I have said the order of battle was that of a hollow square; I should more properly say an oblong square. We had several pieces of artillery, which were judiciously posted. The Red Rovers and New Orleans Greys formed the front line of the square, the Red Rovers being on the extreme right. Col. Fannin took a commanding position, directly in rear of the right flank. Our orders were not to fire until the enemy approached in point blank shot. The cavalry on our right dismounted, about 350 strong, and when within about a quarter of a mile of us, gave a volley with their scopets, which came whizzing over our heads. They still continued to advance, and from the proximity of the second volley of balls to our heads, I ordered my company to sit down, which example was followed by all except the artillerists. The third volley from their pieces wounded the man on my left and several others.

About this time, Col. Fannin had the cock of his rifle shot away by a ball, and another buried in the breech. He was still standing erect, a conspicuous mark, giving orders "not to fire yet" in a calm and decided manner. The enemy had now advanced within about a hundred yards of us; they halted and manifested a determination to give us a regular battle. At this moment, we opened our fire on them—rifles, muskets and artillery. Col. Fannin, at the same time, received a severe wound in the fleshy part of his thigh, the ball passing obliquely over the bone, carrying with it a part of his pocket handkerchief.

At this crisis, the enemy's infantry, from about ten to twelve hundred strong, advanced on our left and rear. Those on our left were the celebrated Tampico Permanent Regiment, of which Santa Anna said "were the best troops in the world." When at a convenient distance, they gave us a volley and charged bayonet. So soon as the smoke cleared away, they were received by a piece of artillery, Duval's riflemen and some other troops, which mowed them down with tremendous slaughter. Their career being thus promptly stopped, they contented themselves with falling down in the grass and occasionally raising up to fire. But whenever they showed their heads, they were taken down by riflemen.

The engagement now became general; and a body of cavalry, from two to three hundred strong, made a demonstration on our rear. They came up in full tilt, with gleaming lances, shouting like Indians. When about sixty yards distant, the whole of the rear division of our little command, together with a piece or two of artillery, loaded with double canister filled with musket balls, opened a tremendous fire upon them, which brought them to a full halt and swept them down by scores. The rest immediately retreated and chose to fight on foot the balance of the day.

Our guns had now become hot—we had no water to sponge them—many of our artillerists had been wounded and we had to rely alone on our small arms. These were industriously handled, as all our men were kept busy during the balance of the day. The action commenced about one o'clock and continued, without intermission, until after sunset. Our whole force did not exceed *two hundred and seventy-five men.* That of the enemy (from all the information we could get) was reckoned at *seven hundred cavalry and twelve hundred infantry!* Our loss was seven killed, besides several mortally wounded and sixty *badly* wounded. We had many others slightly wounded. Out of the number killed, four belonged to my company, and more than one half of my company were struck with balls during the battle.

The courage of all was of that character which would have done honour to veterans. I might particularize many young men whose daring was conspicuous, but from motives of delicacy, I refrain from doing so. My company was more immediately under my view than that of any other. I feel no hesitation in saying, the cool and undaunted courage, the fearless intrepidity and chivalrous bearing of many, very many, would have done honour to Rome and Sparta in their proudest days of military glory.

The enemy's loss was immense, but as we have no correct account of the number, it must be conjectural. Many hundreds must have been killed and wounded. General Rusk has informed me that papers fell into his hands after the Battle of San Jacinto which make the enemy's loss even more than we understood it to be.

Having stated our force at only two hundred and seventy-five men, I deem it proper to give you the names of the companies engaged in the Battle of the Prairie, otherwise called "Fannin's Battle."

Colonel Fannin and Major Wallace.
Red Rovers - Shackelford
Orleans Greys - Pettis
Mustangs - Duval
Mobile Greys - McManeman
Regulars: Artillery - Westover

Capt. Fraser, who likewise commanded the militia of San Patricio, had a few of his men with him. Drs. Barnard and Field were likewise both engaged in battle. The former had the cock of his gun shot away and calmly took a musket from the hands of one of his wounded companions, and resumed his duty with perfect coolness. Capt. F. I. Desauque, the bearer of Gen. Houston's express, was also actively engaged.

Here I mention with much pleasure three other young men: Chadwich, Brooks and Brister. The last mentioned was at the taking of San Antonio, in the first conflict, and was our adjutant. The two former were in Col. Fannin's staff. Chadwich was from Illinois; Brooks from Virginia. They were both gallant and gifted young men. During the battle, Brooks received a severe wound, having a musket ball buried in the centre of his thigh. I afterwards found him at Goliad, in the quarters of some Mexican officers; and the night before the massacre, I extended the limb and dressed his wound. When that horrid scene was passing, this gallant young man was dragged out, in the presence of several Mexican officers, by two soldiers, and put to death with a bayonet.

I have said that Col. Horton had been sent in front to reconnoitre the road about the Coleto; and as much censure has been cast upon this officer by some, for his subsequent conduct, I will relate what I have learned from my second lieutenant, Francis, who was with him, and from one of my company, Joseph Fenner, who was likewise with him. They are both as fearless and gallant fellows as were in the army. They state, that so soon as our firing was heard, Horton ordered all "to horse," having called a halt, and immediately retraced his steps to the edge of the prairie, where they had a view of our engagement, then going on; and from the direction in which the enemy and ourselves were placed, it had very much the appearance of our commingling together, as they saw troops immediately in our front, and others in our rear and on our flanks;—that Horton's lieutenant, Moore, objected to going to our assistance, stating as his belief that we must be cut to pieces; and immediately dashed off, taking the greater part of the force with him: that Horton manifested a willingness to go in; but after nearly all his men had left him, concluded the attempt, with the few men who remained, would be an act of desperation: that they immediately retreated to Victoria, where they expected to unite with a Texan force; but on reaching that place, found that the troops who had been stationed there had retreated; and that a large force of Mexicans was but a few miles off. From the statements of these two men, I did not in the least blame Horton. He might have made the attempt to get in, but I candidly believe that, even with the whole of his force, he could never have cut his way through such an immense number of Mexican cavalry.

During the night, the enemy occupied the strip of woodland in front of us, and we entrenched ourselves on the ground where we fought. It has been often asked, as a matter of surprise, why we did not retreat in the night. A few reasons, I think, ought to satisfy every candid man on this point. During the engagement, our teams had all been killed, wounded or had strayed off, so that we had no possible way of taking off our wounded companions. Those who could have deserted them under such circumstances possess feelings which I shall never envy. I will mention another reason, which may have more weight with some persons, than the one already given. We had been contending for five hours, without intermission, with a force more than *seven times* larger than our own; had driven the enemy from the field with great slaughter; and calculated on a reinforcement in the morning, from Victoria, when we expected to consummate our victory. The morning of the 20th came. But instead of a reinforcement, as we had anticipated, the reverse was the fact. The enemy had an accession to their remaining number of about *five hundred* men.

Their whole force was then displayed in the most imposing and pompous manner, together with about three hundred pack mules, keeping, however, concealed, some pieces of artillery. These, being masked, were placed upon an elevated piece of ground, and were poured upon us, but without any effect. They took care to keep without the range of our rifles. Our cannon had become cool and we could have returned their fire, but perhaps with no effect, and therefore reserved all for close quarters. Here let me remark that I have read Gen. Urrea's pamphlet on this subject, in which he says the firing of artillery was only the signal for the general charge. On this point, as well as his denial of any capitulation, I never read a more villainous falsehood from the pen of any man who aspired to the rank of General.

After they had fired a few rounds at us, they raised a white flag which was soon taken down. We then had a consultation of officers, a majority of whom believed that we could not save our wounded without capitulation; and but *one* solitary man in the ranks would have surrendered at discretion. We then raised a white flag, which was responded to by the enemy. Major Wallace was sent out together with one or two others who spoke the Mexican language. They shortly returned, and reported that the Mexican general could capitulate with the commanding officer only. Col. Fannin, although quite lame, then went out with the flag.

When he was about to leave our lines, the emotions of my mind were intense, and I felt some anxiety to hear the determination of the men. I remarked to him that I would not oppose a surrender, provided we could obtain an *honourable capitulation*, one on which he could rely; that if he could not obtain such, to come back—our graves are already dug—let us all be buried together. To these remarks the men responded in a firm and determined manner; and the Colonel assured us that he would never surrender on any other terms.

He returned a short time thereafter and communicated the substance of an agreement entered into by Gen. Urrea and himself. Col. Holsinger, a German and an engineer in the Mexican service, together with several other officers, then came into our lines to consummate the arrangement. The first words Col. Holsinger uttered after a polite bow were, "Well, gentlemen, in eight days, liberty and home!" I heard this distinctly. The terms of the capitulation were then written in both the English and Mexican languages, and read two or three times by officers who could speak and read both languages. The instruments which embodied the terms of capitulation as agreed upon were then signed and interchanged in the most formal and solemn manner, and were in substance, as follows:

1st: That we should be received and treated as prisoners of war according to the *usages of the most civilized nations.*

2nd: That private property should be respected and restored; that the side arms of the officers should be given up.

3rd: That the men should be sent to Copano, and thence to the United States in eight days, or so soon thereafter as vessels could be procured to take them.

4th: That the officers should be paroled and return to the United States in like manner.

I assert most positively that this capitulation was entered into, without which a surrender never would have been made. I know that when Santa Anna was a prisoner, he flattered many into a belief that no capitulation was made; and those who are disposed to distrust the solemn asseverations of their unfortunate and much injured companions in arms, and take the bare word of an unprincipled tyrant as blood-thirsty as ever foully disgraced the annals of civilization, are welcome to all benefit of such confidence and credulity.

After our arms had been given up and the necessary arrangements made, all who were not so badly wounded as to prevent their marching, were posted off to Goliad under a strong guard. We reached there a little while after sunset and were driven into the church like so many swine. We were compelled to keep a space open in the centre for the guard to pass backward and forward, under the penalty of having it kept open by a discharge of guns. To avoid this, we literally had to lie one upon another. Early in the morning, their soldiery commenced dragging the blankets from our wounded. I resisted an attempt of this sort near me, the result being a bayonet was swiftly drawn and thrust at me.

So soon as it was sufficiently light to see well, I commenced (with what little means I could procure,) dressing and attending our wounded; but I was soon summoned by some Mexican officers, who came to the church door, to attend them. From that moment, I found that I had to labour in the hospital, and that scarcely an hour in the day would be allowed me to attend my wounded companions.

On the second day after our arrival, Col. Fannin and the wounded who were left behind arrived at the Fort; the men having scarcely any water, being compelled to bring it from the river in canteens; nor had we any other food than a scanty pittance of beef without bread or salt. Col. Fannin was then under the protection of Col. Holsinger. On passing from one part of their wounded to another, I made it convenient to see Fannin, and stated to him how badly we were treated. He immediately wrote to Gen. Urrea, adverting to the terms of capitulation and to our treatment. He told me a promise was given him, that every comfort in their power should be provided for us in the future. Let me here ask, if there had been no capitulation, why did not Gen. Urrea advert to the fact, when Col. Fannin urged upon him the immediate observance of its requirements?

The next day, Col. Fannin went in company with Col. Holsinger, on their way to Copano for the purpose of chartering a vessel, then said to be there, to take himself and the men to the United States. When they reached that place, however, the vessel had departed. This, I afterwards learned, was a stratagem to get possession of one of the vessels belonging to Uncle Sam's folks, thinking the old fellow too good-natured to resist any little breach of the kind. On the 23rd, Major Miller and about seventy men were brought in, having been taken at Copano; and on the 25th, Col. Ward and command, taken, as I said before, near Victoria.

Our treatment did not vary much during the week, except that the men were marched into an area of the Fort, without any protection or covering; and the church filled with a part of their wounded (ours occupying the barracks, or rather one room.) On the 26th, Col. Fannin returned. That night I slept in a small room with him and some other officers. This room was in one corner of the church, and was where we kept our medicines, instruments, bandages, &c. Col. Fannin was quite cheerful, and we talked pleasantly of the prospect of reaching the United States.

I cannot, here, resist an inclination to mention one more incident of that evening—the last evening of many, very many gallant spirits. It had a peculiar effect upon my feelings and never can be erased from the tablet of my memory. Many of our young men had a fondness for music and could perform well, particularly on the flute. In passing by them to visit some wounded, on the outside of the Fort, my ear caught the sound of music, as it rolled in harmonious numbers from several flutes in concert. The tune was "Home, Sweet Home." I stopped for a few moments and gazed upon my companions with an intense and painful interest. As those "notes of mournful touch" stole upon the breeze, the big tear that rolled down many a manly cheek, which had glowed in battle and burned in the rage of conflict, told the heart's irrepressible emotion; for the image of home and friends came over the mind "like the pressure of a spirit-hand."

Poor fellows! It was their last earthly evening. Little did they dream that the next morning, *Treachery* would consign them to their everlasting home! Subsequent events rendered it easier for me to forget all scenes of a thousand days of pleasurable enjoyment, than to cease to remember this one incident of those few lonely minutes of grief.

27th March—Palm Sunday—Never whilst the current of life rushes through this poor heart of mine can I forget the horrors of this fatal morning. At dawn of day we were awakened by a Mexican officer calling us up, and saying he "wanted the men to form a line, that they might be counted." On hearing this, my impression was, that in all probability, some poor fellows had made their escape during the night. After leaving the church, I was met by Col. Guerrear, said to be the Adjutant General of the Mexican army. This officer spoke the English language as fluently as I did myself; and to his honour be it said, he seemed a gentleman and a man of feeling. He requested that I would go to his tent in company with Major Miller and his men; and that

I would take my friend and companion, Dr. Joseph H. Bernard with me.

We accordingly went over to his tent, about one hundred yards off, in a southwesterly direction. On passing the gate of the Fort, I saw Ward's men in line, with their knapsacks on. I inquired of them where they were going. Some of them stated that they were to march to Copano, and from thence be sent *home!* After reaching Col. Guerrear's tent (to attend to some wounded, as we expected,) we sat down and engaged in familiar conversation with a little Mexican officer who had been educated at Bardstown, Kentucky. In about half an hour, we heard the report of a volley of small arms, towards the river and to the east of the Fort. I immediately inquired the cause of the firing, and was assured by the officer that he "did not know but expected it was the guard firing off their guns."

In about fifteen or twenty minutes thereafter, another such volley was fired, directly south of us, and in front. At the same time, I could distinguish the heads of some of the men through the boughs of some peach trees, and could hear their screams. It was then, for the first time, the awful conviction seized upon our minds—that *Treachery* and *Murder* had begun their work. Shortly afterwards, Col. Guerrear appeared at the mouth of the tent. I asked him if it could be possible they were murdering our men? He replied that "it was so"—but that he "had not given the order; neither did he execute it." He further said he had done all in his power to save as many as he could, and that if he could have saved more, he would have done so.

The men were taken out in four divisions, and under different pretexts; such as, making room in the Fort for the reception of Santa Anna—going out to slaughter beef—and being marched off to Copano to be sent home. In about an hour, the closing scene of this base and treacherous tragedy was acted in the Fort; and the cold-blooded murder of all the wounded, who were unable to march out, was its infernal catastrophe. I learned from the interpreter that Col. Fannin was the last doomed captive of vengeance. He was ordered to communicate the fact to him and Fannin met his fate in a calm and soldier-like manner: that he handed his watch to the officer who superintended his murder, with a request that he would have him decently interred; and that he should be shot in the breast, and not in the head; with all of which the officer solemnly promised to comply; that Fannin was then placed in a chair, tied the handkerchief over his eyes with his own hands, and then opened his bosom to receive their balls. Major Miller, who knew Fannin, informed me that the next day he saw Fannin lying in the prairie among a heap of wounded, and that he was shot in the *head*!

We marched into the Fort about 11 o'clock and ordered to the hospital.—Had to pass close by our butchered companions, who were stripped of their clothes, and their naked, mangled bodies thrown in a pile. The wounded were hauled out in carts that evening; and some brush thrown over the different piles, with a view of burning their bodies. A few days afterwards, I accompanied Major Miller to the spot where lay those who were dear to me whilst living; and whose memory will be embalmed in my affection, until this poor heart itself shall be cold in death;—and Oh! what a spectacle! The flesh had been burned from off the bodies, but many hands and feet were yet unscathed. I could recognize no one. The bones were all still knit together, and the vultures were feeding upon those limbs which, one week before, actively played in battle.

I will here relate an incident which I received from the lips of one of my company who made his escape. When the division of the army to which he belonged was brought out and made ready for the work of destruction, the men were ordered to sit down with their backs to the guard. Young Fenner (the same who had shot the spy glass from the hands of the officer, as before mentioned), rose on his feet and exclaimed, "Boys, they are going to kill us—die with your faces to them, like men!" At the same moment, two other young men, flourishing their caps over their heads, shouted at the top of their voices, "Hurrah for Texas!" Can

Texas cease to cherish the memory of those whose dying words gave a pledge of devotion to her cause?

Many attempted to escape, but were run down by the cavalry or shot; and considering the nature of the place of butchery, with all the difficulties by which they were surrounded, it seems like a miracle that even a solitary one should have succeeded; and yet some did escape. From all the information I could get, from as many of these as I have conversed with, and from other sources, I herewith subjoin the names of those who escaped, and also of the companies to which they belonged:

Orleans Greys—Wm. L. Hunter, Wm. Brannan, Jno. Reece, David Jones, B. H. Holland.

Huntsville Volunteers—Bennet Butler, Milton Irish.

Mustangs—Wm. Morer, Jno. C. Duval, Wm. Mason, Jno. Holliday, Jno. Van Bibber, Charles Spain, —— Sharpe.

Burkey's Company—Herman Eremby [Ehrenberg], Thos. Kemp, N. J. Devany.

Horton's Company—Daniel Martindale, Wm. Hadden, Charles Smith.

Red Rovers—Isaac D. Hamilton, D. Cooper, L. M. Brooks, Wm. [Wilson] Simpson.

Company not recollected—N. Hosen, Wm. Murphy, Jno. Williams.

The physicians who were retained were Dr. Joseph H. Bernard of the Red Rovers, who had been appointed by Col. Fannin as surgeon to the garrison; Dr. Field, who had been sent on by the government; Dr. Hall, who was a member of Major Miller's command; and myself, whose professional services, as before stated, had been called into requisition by the Mexicans. We had previously detailed several men from the ranks as assistants in the hospital; of the number, Bills, Smith, Griffin, and Skerlock were alone left. Our situation at this time was truly deplorable; having everything stolen from us but the clothes on our backs, without any of the comforts of life;—compelled to labor in the hospital day and night;—exposed to a piercing March wind, and no blankets to cover us during the night;—having little or no food, and that of the most revolting kind;—covered with vermin, worn down with fatigue, and prey to the most heartrending forebodings. But still under every discouragement, we sustained ourselves with becoming fortitude. The officers, in their intercourse with us, evinced great politeness, which they seemed to consider the sum total of their duty. On one occasion, they invited Dr. Bernard and myself to eat with them, but we gave them such a demonstration of American appetite as to admonish them in the future of the bad policy of extending to us such acts of hospitality.

I consider it not inappropriate here to mention one female, Pacheta Alevesco, the wife of Captain Alevesco. She was indeed an angel of mercy—a second Pocahontas. All that she could do to administer to our comfort—to pour "oil into our wounds," was done. She had likewise been to Major Miller and men, a "ministering angel."

Our regular routine of service was kept up until about the 20th of April [Barnard's narrative suggests a departure date of April 16th], when an express arrived from the Alamo, requesting the attendance of some physician who could amputate a limb, as many had died at that place for the want of some person possessing that skill. My friend, Dr. Bernard, was selected for that duty, and I gladly availed myself of Col. Ugartachea's consent to accompany him. Horses were provided for us, and we set off under the guard of a sergeant and private of cavalry.

I have been often asked, why did we not then make our escape? One answer will suffice. These men were extremely kind and attentive to us; seemed to repose perfect confidence in our integrity and honour—had not been engaged in battle against us; and had done us no injury. The chance of escape, without taking their lives, was doubtful;—to do such a deed in cold blood was what we revolted at.

We arrived at San Antonio on the fourth evening after leaving Goliad. I was immediately conducted to the headquarters of

Gen. Andrada. I found him wearing a fine fur cap, which had been the property of some of our gallant countrymen who had fallen in the Alamo; smoking a cigar held by golden tongs; and surrounded by some of his principal officers. From the immense crowd of men, women and children that followed us, it was evident we were the objects of as much interest as a caravan of wild animals would be, if led through one of our principal cities.

The general received us politely, read the letters which had been sent with us, conversed in French with Dr. Bernard, and promised us our passports, so soon as some few who were badly wounded should begin to do well. These passports were never given us. Dr. Bernard was taken to the house of Don Navarro; and I was conducted to that of Don Ramon Musques. We were politely received and kindly treated by them and their families. I say this with much pleasure, as it awards to those families a tribute justly due them. A new era in our destiny seemed now to open upon our astonished senses. We were here disencumbered of the cannibals which had been preying upon our poor carcases for the last four or five weeks; obtained, if not decent, at least clean clothes, and met with smiling and pleasant faces in officers, men and inhabitants generally.

Our duties were confined to the hospital, but we had assistants, and they were not of a very arduous character—found about four hundred wounded men at this place—commenced practicing medicine among the inhabitants, although the pay was of a very low grade. We remained here until the Battle of San Jacinto. That event seemed, at first, to throw no little consternation into their ranks, but it was soon forgotten. We applied to the General for passports but could not get them. The army made preparation to leave and join Filisola on his retreat. No confidence seemed to be placed in the arrangement made by Santa Anna with the Cabinet of Texas. One company was left behind, and the care of the wounded was committed to us.

Two days after the army departed, we put off, early in the morning, having first provided ourselves each with a good horse, guns, pistols and ammunition. These arms were procured in a manner that would not have been deemed proper under other circumstances. We made directly for Goliad, passed the army of Andrada in the night and kept clear of the road until the third day. We are greatly indebted to Dr. Alsbury and family for their friendly aid in this matter.

The third morning, we met in the road an officer of the lancers and six men. They were within a short distance of us before we discovered them; and we were fearful of a rencontre against such odds, but rode directly up to them and, by a little address, "threw sand into their eyes," and passed on.—Reached Goliad next morning and found there about fifteen Texan troops. Here we rode over the ground which had drunk the heart's blood of our mangled companions. Their bones were bleaching on the prairie. The rage of battle had passed and all was calm in the stillness of death. Imagine our sensations. "When the tumult of battle is past, the soul in silence melts away for the dead."

That night we met a part of Gen. Rusk's army about five miles from Goliad and encamped with them. The next morning, we met Gen. Rusk and requested of him, as a last sad duty, that he would have the bones of our fallen companions interred with the honours of war. This he promised, and faithfully performed. We then went on to Velasco, where we found the Vice President, Zavala, President Burnet and many other distinguished men. Here, too, was that fiend incarnate, Santa Anna, then a prisoner, whose deeds had been called aloud to heaven for vengeance and just retribution. He was treated with that kindness which should have been shown him by a gallant, but unfortunate foe. Genius of Fannin! come forth and confront the lying dastard! Murdered companions of Goliad! call on your country for vengeance!

Here I obtained an honourable discharge and hurried home to console a bereaved and disconsolate family, for the loss of a son, a brother and others who were dear to them. [Shackelford lost his eldest son, Fortunatus Shackelford, a nephew, William J. Shackelford, and a former student in the Massacre.] My friend,

Bernard, remained behind. This gallant fellow, who had nobly fought by my side, who had been my companion in every trial and difficulty, determined to stay and aid the cause he had espoused, with his last dying effort.

On reaching home, I found, so strong was the universal conviction produced by the report of my death, that I had been buried with honours of war, together with most of my company; and my life and deeds and last days had passed in review before a multitude, which assembled to hear a funeral sermon that was pronounced upon the mournful occasion.

Abel Morgan, a.k.a. Thomas Smith

[*When Abel Morgan joined the Texian army in the fall of 1835, according to his 1870 pension claim, he entered under the name Thomas Smith. He was honorably discharged, then enlisted under Ira Westover under the same name in early 1836. It was under his Smith pseudonym that he participated in the Battle of Coleto Creek. Morgan was saved from the ensuing massacre by acting as an orderly under Dr. Shackelford after the battle. In Kentucky in 1847, he published a pamphlet detailing his experiences at Goliad. The text that follows is extracted from that publication.*]

As I have received many solicitations from my friends who live at a great distance from me to give them a written account of the battle and Massacre at or near Goliad, which is also called Labide; and as my mental faculties appear to be in a better condition at present than they have been since the fatal event, I feel it my duty to state what I can now remember concerning that transaction.

Some time in March 1836, perhaps about the 10th, Col. Fanning received an express from Gen. Houston, ordering him to evacuate the fort at Goliad and come on to him. But it was reported that Col. Fanning said he would take the liberty to disobey the order, and risk a battle, as he had something like 500 men at that time. A few days afterwards he received an express from St. Antonia de Bexar from Col. Travis, requesting some help from him. I suppose about 200 of our men started and crossed the river and struck their camp in sight of us remaining there until the next evening and then returned to us again within the fort.

By this time Col. Fanning had concluded to obey orders and leave the fort, when he received a petition from a man by the name of Ayes who lived at the Mission, perhaps 20 miles off, to send a guard to escort him and his family to Goliad as he wished to go with us. Captain King went with his company. Arriving there, he found about 200 Mexican soldiers. He sent back for help. Major Ward went to assist him with the Georgia battalion, and they had a considerable battle. Ward made out to retreat with his men, but the Mexicans drove King and his company into an old Church. There King and his men bravely fought until they were all killed. Major Ward retreated into the Warloop [Guadalupe] River bottom. Two days after Fanning's capitulation with the Mexicans they took Ward and his men and put them into prison with us. This left us but about 360 men.

I think it was on the 19th of March in the evening, there was a smoke discovered about two miles up the river on the north side in the timber. We had but about 40 horsemen. They were volunteers from Matagorda sent out to see what discovery they could make. After a little while they came dashing back pursued by a considerable force of Mexican horsemen. Our men took shelter in an old Church on the opposite side of the river from us, about six or seven hundred yards from us. The men in the fort then turned out to their assistance to the number of 200. The Mexicans then fled, and our brave little squad of horsemen pursued them. When our men would turn to come back the Mexicans would pursue them until they would get within gun shot of our footmen, when they would turn and our men they pursue again. They kept alternately chasing and being chased until dusk when the Mexicans left. Our men returned into the fort, all having escaped without injury. What damage was done to the Mexicans we never learned.

Previous to this time, for about a month, I had been kept in the hospital and had charge of three sick and one wounded man. The wounded man was Solomon Hamilton from Mississippi. (He was shot and wounded by one of our guard in the night, because he did not answer the guard's challenge.) The sick men were John McGowan from Alabama who had the dropsy, McCoy from Mobile, who had consumption, and Debusk from Alabama. I had a young man to help me take care of these men, but that night, after this skirmish was over, the officer of the guard came to the room that we made us into a hospital and told me that I was requested to stand guard that night and to take the first watch on the northeast bastion.

Accordingly I went and took my station. I suppose in about half an hour Col. Fanning and Capt. Westover came to me, Col. Fanning asking me what I thought about retreating and leaving the fort. I told him that my opinion was that is was too late, for I made no doubt from what we had seen that we were entirely surrounded by the enemy, and that we had something like six weeks' provisions and men enough to keep the enemy from breaking in for some time, as we had then about 360 men. Col. Fanning seemed to have his mind unsettled about it. Capt. Westover agreed with me, and said if we had left some three or four days before, he thought we might have escaped, but he made no doubt that we were surrounded now. Capt. Westover was captain of the Regulars, and I belonged at that time to his company. I had served as a volunteer four years previously. Capt. Westover told me he thought that if they attacked us that night that they would be very apt to attack the place where I was, as there was a kind of a gully leading from the river to that point. I told him if they undertook it while I was there I should turn loose the cannon on them for an alarm gun. He said it might take too much time to get the match. I told him that I did not want the match. I could turn down the apron of the cannon, stand aside and turn the lock of my musket in it and fire it off. He answered, "Give me you yet," and they left me. When my tour was out I went to the room where I used to mess before I was put into the hospital. All hands were busy, and stated that we were to start in the morning to leave the fort.

In the morning it took until about 8 or 9 o'clock to get breakfast and to destroy our stock of provisions. When we got to the ford of the river, our largest cannon got into the river and we were detained, perhaps an hour. We had to haul our cannon with oxen and they were wild and contrary, and by the time we had gone three or four miles we had to stop and rest them. We lost some time in this way when we got about seven miles from Goliad we entered a prairie perhaps from three to five miles across. By the time that we got about one mile into the prairie, the whole western border of the prairie was lined with Mexicans, and by the time that we got half a mile further they broke in a cloud, as it were, ahead of us to the east. We had nine cannon of different sizes along. We halted and fired several rounds with one of the cannon and then geared in again and went on perhaps a mile further.

By this time the Mexicans had surrounded us except one little gap to the south, then we were ordered to halt and prepare for a general battle. Then our few horsemen left us with the expectation of returning with a reinforcement from [Victoria] on the Warloop [Guadalupe] river which was nearby. But the Texas forces that were stationed there had left before they got there as we supposed, as they never returned to us.

There was immediately a square formed, and as the oxen were taken from the cannon instead of being secured, they were turned loose and got away and they went right off to the Mexicans. I had my sick men and the cripple in a wagon. The four men and their sacks filled it. I had two yoke of oxen and was the foremost team. So when they halted and formed a square, I was left forty or fifty yards from the square to the east. On the way one of the officers told me to give him my gun and he would put it in the baggage wagon for it would be as much as I could do to attend to the sick.

There was a Mexican employed to drive the oxen and, as soon as he was ordered to halt, he let the wild cattle go and ran off to the [enemy.] I was left to head the oxen or let them carry the sick men right into the enemy's ranks. About the time that I got ahead of the oxen, one of them received a ball that stopped the wagon. Soon after, another one of the oxen was crippled so bad that I knew my wagon was safe. I looked around and found there was no chance for me to get water for the sick ones and saw that I could do them no further service. I walked into the square. I knew we had some new muskets in the ammunition wagon. I selected me one of them and catched up two packs of cartridges and walked out to my wagon again where the balls were whizzing about like bees swarming.

About that time after I had fired eight or ten times by myself, there came out four more men, and we formed a platoon of five. An Irishman by the name of Cash was at the head of the platoon. I was next. A Dutchman by the name of Baker next. A young fellow from Georgia next. A man by the name of Hews next. The last had a rifle and I suppose he is alive yet as he escaped the Massacre. He was a Georgian. He was not a soldier but a visitor and had his horse and gig along. After a few rounds Cash received a ball in the corner of his head and as he fell he handed me his gun, saying, "Take this; she won't snap." My gun had got muckey with powder, and missed fire, and he had noticed it. I took [his gun] and kept it the balance of the day. The ball cut the size of it out of his head but did not kill him. In a short time Baker, who stood at my left hand, was shot down. He had his thigh broken and before he was carried into the square he got another ball in the body. The both were carried into the square, and Hews took advantage of the wagon to rest his rifle on. There was a low tree, from 140 to 170 yards distant where the Mexicans would creep up and shoot at us. Hews killed two and wounded a third at that tree. I went and looked at them next morning.

The young Georgian and myself were left by ourselves. The Mexicans came up tolerably close and there was a fine dressed fellow maneuvering as if urging his men on mightily. We both agreed to take aim at his head, as soon as he would halt. We did so, or at least I know I did, and the fellow tumbled from his horse. His men turned and retreated some distance. This daring fellow from Georgia laid his musket against the wagon and run and took from the dead Mexican a purse of money about seven or eight inches long. He came up shaking it. I told him that was a foolhardy trick, for the Mexicans were mounted and if they had looked behind and discovered him they would have turned upon him and cut him to pieces. He then took his musket and went to the square.

After he went to the square another rifleman came out to the wagon and stayed with me a while. I think it was Doctor Bernard. If it was, I hope he is still alive. He is one of the four doctors who were saved at the Massacre. I then went into the square after some more cartridges and to see if Cash was dead, but he had revived as the film of the brain was not broken. I was much mortified to see so many fine fellows laid down there with their blankets spread over them.

At this time four of our cannon were idle because the regulars were wounded or killed, and Capt. Westover said that the volunteers did not like to undertake to man them, and allowed that they could do as much good with their rifles. I suppose it was something like 1 o'clock when the square was formed and we began to turn loose our cannon on them. We kept that place in a continual tremor. From that time until after sundown our men appeared to be just as much composed and as busy as they had been at no other work, but it looked like a great odds to see 360 men in a little band surrounded by something like 1,900, and 7 or 800 on horseback. But many were the Mexicans I saw leave their horses that day who never were to mount them again. It was generally supposed that we killed about 200 Mexicans that day.

After sundown the Mexicans quit firing, retired to a distance of about a mile, and struck camp in the edge of the timber. We went to work ditching. Every animal we had was killed or wounded except two of the oxen that I had in the wagon. We took the dead mules and horses, and laid round and made breastworks by ditching and throwing dirt on them. Even our knapsacks were piled on to help, and some trunks. We soon had one square fortified and then we had to look to our dead and wounded. We had nine men killed and fifty-one wounded, besides Col. Fanning who received a ball in the abdomen near his hip in the early part of the engagement. He did not give up for it, but kept about most of the time. Capt. Westover, Doctor Shackelford, and several other officers walked around the square and attended to their business in a brave and honorable manner. Doctor Shackelford was captain of the Alabama volunteers. He was one of the doctors who were saved at the Massacre. I hope he will live to forget the trouble he saw there, for he is a brave, high-minded, noble-hearted man.

We had twelve or fifteen Mexican prisoners with us. So soon as the square was formed they got bayonets and began to dig holes in the ground and soon let themselves down underground and so escaped being hurt. My reason for mentioning this is that I have been twice asked by men who have seen our battleground what we dug those round holes for. This will explain why the holes were dug.

As soon as dark came on the whole prairie resounded with the sound of bugles and they kept it up all night. After night a while there was a vote taken whether we should retreat and leave our wounded or stay with them. I soon put the question to myself—if I had an arm or leg broken and my comrades were to leave me defenseless that I never could forgive them. I answered that if we could not manage to take the wounded men with us that I was for staying with them. We had a large majority on that side of the question. We then prepared ourselves to guard against cavalry. Many of us had two muskets. We were down on one knee behind our little embankment with one musket in our hands and braced against our knees and the other leaning against the embankment. Our orders were if they came to charge on us that night, to let them come close enough so as to be sure to kill one out of each fire, for every man, and then to use the bayonet.

We remained in the position all night, and the Mexicans continued to blow bugles. About midnight we saw and heard some persons on the north side of the square perhaps 100 yards or more off, but they came no nearer. It was a misty night and we could not see very far. The next morning early they fired a chain shot out of a large cannon at us which made a wonderful whizzing over our head. Directly they fired another and then quit, and hoisted a white flag. Col. Fanning had one hoisted, and the interpreters passed and repassed. At length a Mexican colonel met Col. Fanning, and they made a capitulation, to which Gen. Urrea, who was commander of the Mexican forces at that place, agreed.

I asked Col. Fanning upon what conditions he was to surrender his arms. He told me that our lives and our private property were guaranteed to us; that we were to give up to the Mexicans all of the Texas property that was in our possession, and that as soon as they got their provisions out of their vessels that were at Copino [Copano], that all of us who belonged in the States were to be sent to New Orleans; and that what few prisoners they had, that belonged to Texas they were to retain as prisoners of war until exchanged. That, I think, is about what he told me.

As soon as they were done writing, we marched out and laid our muskets in one pile, and then marched a little further and laid our pistols and durks. I laid down a pair of nice brass barrel pistols that I had belted around me, and as I drew out my butcher knife, I opened my mouth and showed the officer that I had but two or three teeth and motioned to him that I could not eat without the knife. Then he motioned to me to keep it.

Our ammunition wagon was in about the centre of the square and while we had been fighting the day before there was hundreds of cartridges dropped and the dry grass was very high and it got trampled down and hid the cartridges. As soon as we had given up the square to the Mexicans, they crowded in with their segars in their mouths and directly the dry grass took fire and the wind sprung up and in about five minutes the whole square was in a flame. Every moment there was a volley of cartridges bursting. About this time you might have seen Mexicans and Americans at the top of their speed leaving the square, expecting the wagon to blow up. But it did not take fire. There were some few of our men still in the square. Two got badly burned by the bursting of the cartridges. Of all the mad people ever I saw the Mexicans were the most enraged, for they thought that we had done it on purpose.

After the business was arranged we were started back to Goliad. Now we had not had a drink of water or a morsel to eat from Sunday morning, and this was Monday evening. About dusk we got the chance to drink, for we had to wade the Stanton [San Antonio] River about armpit deep, and then marched up to the Church dripping wet to sit flat down on a stone floor. One sat and leaned on your back, you leaned on another's back. There we sat until next evening. We had then been without food from Sunday morning until Tuesday evening. We were taken out of the Church and put into what might be called a pen or certain boundary with guards all around. This evening we were to draw rations. I got a bit of fresh beef, boiled it in my tin cup, and when it was done it was a not bit bigger than a turkey egg. I had no salt, no bread—nothing but to eat the beef and drink the broth from it. That was the first I had eaten from Sunday morning. I cannot say that I suffered with hunger. About that quantity was what we all got and but once a day until after the Massacre.

About this time the Mexicans took Major Ward and his men and put them in prison with us. Major Miller landed about the same time at Copino with about 80 men. The Mexican officers took Col. Fanning down to Copino to get Miller to surrender. I do not know what arrangement was made but Miller and his men came up to Goliad and were stationed out about a quarter of a mile from us and had the privilege of working about for themselves.

I think our wounded men were hauled in from the battle-ground on Tuesday evening, to Goliad, and were put into the old Church. Twelve of us were put in there to wait on them. We had 51 wounded men and the four that I had before in the wagon made 55 we had to wait on. On the next day the Mexicans hauled in their wounded. How many there were I cannot say, but they had two hospitals outside of the fort and they placed 57 in the Church where ours were. The wounded Americans filled one side of the Church and the Mexicans filled the other. Our men lay quietly and it was seldom that you would hear them complain, but as soon as the Mexicans came we had music enough. To me it was tiresome, for there was no end to it day or night. The Mexican officers would not let the Mexicans who lived there sell our men anything to eat if they saw them. What they sold was at an extravagant rate. I saw them give four bits for a tortilla, a little corn cake not larger than the top of a saucer, and not as thick as a knife blade. A hungry man would have eaten a dozen of them and then not have had enough. We had to make soup in a large copper boiler for all the wounded. I was attending to it one day and a Mexican came to the cooking place with a calf that had been taken out of a beef cow. He hung it up and skinned it, and cut off one quarter, and then motioned to me to put the other quarters into the soup, which I did. I cut the feet off and threw them down close by the kettle and some of our men came and picked up those feet and roasted them and ate them hide and all.

I think we lived at this starving rate until the next Sunday. I think our battle was on Sunday and that the Massacre was on the next Sunday. Early in the morning the word was, that all of our men but a few, were to be sent down to Copino. Just enough were to be left to nurse the wounded and the rest were to go to the Copino to build a fort. The men put on their knapsacks and were mustered outside. I suppose that there was something like three hundred of them taken out at that time. We who were attending the wounded were ordered to take them out and lay them on the yard before the door. They said they were going to put them in another room. As I was passing through the room where they lay I saw one of them sitting up and saw the tears dropping off his cheeks. I asked him what was the matter. He said those men they have carried out are to be shot, and we all will be shot in an hour. I went out and told the rest of the nurses that we had as well cease carrying out the men, for we were all to be shot. We then quit.

About that time, we heard the guns. Directly the Mexicans came to the hospital and took away ten of the nurses and left only myself and a man by the name of Wm. Shirlock. Directly an officer came and took Shirlock and me to the calaboose, and motioned to us to hide behind the door shutters. The shutters were very wide but they were pushed so close back against the wall that when Shirlock stepped in first and hid, the shutters would not quite hide me. The man who put us there went into town. After a little while the Mexicans came for the rest of our men, except the wounded. They formed a line in front of the calaboose.

As an officer passed, he saw me and he stormed out and motioned to me to come to him. I saw at once that there was no chance for me and I sprang up and went to him for fear that as he came after me he might see Shirlock. I was placed at the end of the row. As they were about starting off, the officer who put me behind the door shutters happened to be passing into the fort and saw me. I suppose he knew me by my hat for I had a hat made in Louisiana that was larger than common and of the natural colour of the fur, so that it was different from the hats worn by my comrades. In passing he turned around and talked awhile with the other officer. Then he walked up to me and gave me a push and motioned to me to go back behind the shutters again. I did so. I suppose we stayed there about one hour and a half.

We heard the guns firing in the fort where they were killing the others. We did not see any of it. After they were done killing them, they came into the calaboose. It was a long building and there came in as many as could get in there, and sat down flat on the floor, which was made of brick. There they laid down the bloody clothes and knapsacks that they had taken from our dead men, and they put a kind of seat for me and Shirlock and took us from behind the door and put us to sit facing them. The knapsack that was in front of me had the name of Wingate on it. There was a pile of boards in the calaboose and when one of our men by the name of Voss, who understood Spanish, found out that the Mexicans were about to shoot us, he hid himself under the boards. At this time, he came out. The [Mexicans] cursed awhile at him and then brought him and sat him by us. There was some soup brought to the soldiers in little wooden bowls.

They commenced eating. But I noticed several officers who could not eat, but were shedding tears by which I was convinced that there were some human beings among these savages.

Near where we sat there was a little boy eating. He was about seven or eight years old. He would take a sup or two and look up at us. At length he got up and stepped to me and offered me his bowl of soup. I shook my head, but he would not desist until I took three spoonsful of it and handed it to Shirlock, who took one spoonful and shook his head. We handed it back to [the child.] The little fellow pointed at our foreheads and shook his head to try to let us know that we were not to be shot.

After they were done eating they brought an interpreter and told me and Shirlock that we were saved for the purpose of waiting on the doctors that they had saved to doctor their wounded. Voss was given to an officer to wait on him. They then brought some bleached domestic and tore it into strips about two inches wide and about two feet long, and tied them round our left arms. They told us that if we lost them, the first soldier who saw us would kill us. Then they took us through the yard of the calaboose to the old Church, and as we passed along I saw our poor men laying dead, stripped of all their clothes. They were throwing them into a wagon to haul them out of the fort. When we got to the Church we found some more of our poor comrades here still alive, saved to wait on the Mexicans.

They fixed up a barrel, that I suppose would hold about forty gallons, with ropes and a pole. Shirlock and I were sent down to the river from 200 to 500 yards to pack a barrel of water. When we got there I suppose there were from 200 to 300 Mexicans strung along the river bank washing the clothes that they had taken from our murdered men. The edge of the river, where it was about ankle deep, was red with blood that had come out of the clothes of the dead men. We waded in about knee deep before we put the barrel down to get our water. While we were getting our water they were making signs that we ought to be killed.

From that time Shirlock and I had to pack water for all in the fort—to undo the bandages of all their wounded, wash their wounds and tie on the bandages, as the doctors would put on the plasters, and at night he or I had to be always up to wait on them. What time we got to sleep was by turns, for we had but one blanket and had to lay on a stone floor, among wounded Mexicans. They smelled as bad as if they were dead already, for in spite of all that could be done for them, the wounds were full of large worms. They were all allowed to lie out in the prairie until their wounds were fly-blown before they were hauled in.

I will name those that were saved. All that I can remember. The four doctors were Dr. Shackelford, Dr. Bernard, Dr. Fields and Dr. Hale. There was one interpreter saved named Spawn [Spohn]. Hews, the rifleman, and Pete Griffin. Griffin bargained with a Mexican officer to save Hews for a large sum of money. Pete could talk Spanish and the officer had to save Pete also. Besides these I have mentioned there was an Irishman named Phagin [Fagan], a blacksmith, two dutchmen who were carpenters (I do not know the Germans' names) and the young man who told me we were to be shot, whose name was Andrew Boyle. He was one of Westover's men, and messed with me before I was placed in our hospital.

Perhaps Doctor Shackelford had as hard a trial as any man there. He had a son, a young physician just grown, a nephew, and a student all shot. None of Major Miller's men were shot. But our men who fought in the prairie, and Ward's men made together about 420 men. All these were shot except fourteen or fifteen. A few however, who were carried out and shot at, made their escape. Dr. Shackelford wrote to me after I got home that three of his men who were to be shot, got away and reached home before he did.

Shirlock and myself had to pack wood as well as water. The interpreter ordered us to go and pull up pickets from the Mexican fences and sometimes, when we would get almost to the fort with a load, a Mexican would meet us and make us carry it back. In one of our hunts after wood we got out to where one division of our men were shot. I suppose from appearances that about one hundred were shot at that place. Their bones were mostly hanging together, though the flesh appeared to be entirely gone. We looked at them in silence a while, and then walked away.

Six or eight days after our men were shot, Santa Anna sent orders for the Mexicans to shoot Miller's men and those of our men who were saved from the Massacre. I believe that the Mexican soldiers were as glad of it as if they had received a great bounty, for they were all singing and rubbing up their guns, and very full of fun and frolic. Indeed we were all amazed for we knew by their maneuvers that there was something at hand. Shirlock and I went down to the river after water and Shirlock observed to me, "Old man, they are going to make a clean turn of us in the morning." About that time I felt as if it would not make much difference with me, for we were kept at work day and night and if we could have had time to sleep, who could have slept while there were hundreds of wolves and dogs eating the remains of our fellow soldiers, in our hearing?

The next morning another courier came from Bexar with orders, countermanding the bloody directions received the day before. Then the Mexican officers told our interpreter to tell us all about it and that we need not be uneasy. About this time Dr. Fields was missing. Whether he got away or was killed by some of the Mexicans I cannot say. The report was both ways. Three of the doctors were now gone. The Mexicans had observed that when I had finished washing the wounds I would assist our doctors in spreading plasters and putting them on so they said I was a doctor, and gave me fourteen of their wounded men to attend to altogether. By this arrangement I got clear of packing water, wood, the soup kettle, and all.

I had not been acting as doctor very long before one of my patients took the cholic. I asked Pete what ailed him that he kept such a fuss. Pete thought he had the cholic. I went to the medicine chest and got a bottle of camphor and rubbed the sick man's stomach with it. In a little time he was speechless, and presently two of the Mexican doctors came in and began talking at a mighty rate. I asked Pete what they were going on so for. He said I would soon find out if that fellow died, for they would have me shot. Pete stated further that my patients had told their doctors that I had a bottle there and that the doctors suspected that I had given him something to kill him.

The fellow lay insensible about half an hour, and then revived. One of the guard went and brought the same doctors, and they questioned him closely. I got Pete to attend and listen to them. The fellow told them that I had not given him anything to take, but had rubbed his stomach with something without which he would have died. They were well pleased and said they knew all the time that I was a doctor, but that I did not want to let them know it. The Mexican got his knapsack and took out a dollar for me. I did not care much to take it but he told Pete that I must accept so as a compensation for saving his life, and that he knew it would get me some sugar or something to eat. But I made up my mind afterwards that if they all died with cholic, I would do no more doctoring for them, further than to dress their wounds. I was very near being shot at that time, for the old fellow did not live long.

Not long after, they came near shooting Pete Griffin. He had all the labor of washing bandages to perform, except what little help he received from Shirlock, who sometimes aided him when he was not otherwise engaged. They worried Pete and made him so mad that he cursed the whole of the Mexican officers and soldiers for a set of blood-thirsty savages, to their faces, in the Spanish language. They took him with the avowed intention of shooting him. Pete defied them, for he felt that his condition could not be made worse. This is what Pete told us afterwards, for neither Shirlock nor myself understood three words of Spanish at that time. They put him into the calaboose three or four days and then took him out again and set him at his old task of washing the bandages.

Massacre

At the time of the Massacre they took all of our clothes from us. The interpreter told us that the Mexican physicians had said that if we could find any of our clothes we should have them. I went to searching and found one of my shirts and one blanket. I do not think that any of the rest got anything at all. Mine were marked. Pete had not a change of clothes and his employment caused him to be very offensive, as he was careless about keeping himself clean.

So soon as the Mexicans heard that General Houston had taken Santa Anna, they piled up the skeletons of our men and made a fire with their fences, and burned up the bones. About this time some of Miller's men were at work near the road that leads from the Warloop [Guadalupe] to the Nueces River, to St. Patricio, and they saw a family of Irish, and they told Miller's men that General Houston had captured Santa Anna and the greater part of his army. They came in and brought us the good news. They went into a grog shop and got to drinking, and began to boast about this occurrence. The Mexicans, I suppose, were frightened, for they sent round and got all the rancheros to come in. They told our men that the rest of Santa Anna's men had overtaken Houston, and had taken Santa Anna and Houston and all his army. That they intended to bring Houston and his army here to be shot. Three large church bells were placed in the fort, and the Mexicans commenced firing guns and ringing bells with desperate enthusiasm. I heard as much racket that evening as ever I did in the same length of time in my life.

When the noise began I had just bought a little parched coffee from a soldier and was sitting on a block in the church preparing my coffee with handle of my knife and a tin cup. When I had put my tin on the fire with my coffee I began to look about for my comrades. Dr. Hale, Shirlock and Pete were all gone. I went to the door and looked out but I could not see one of our men or Miller's men. I began to think they were shot.

The bells and guns kept a wonderful racket. I walked back to where my coffee cup was on the fire and commenced blowing the coals. One of the soldiers who was standing guard at the door quit his post and came to where I was. I had taken my hat off and laid it beside me, and I had a very good silk handkerchief in it. He took my handkerchief out and motioned to me that I was to be shot and that he was going to take my handkerchief. I laid hold of him and took my handkerchief and put it back on my hat and motioned to him to go away. He pointed his finger to his forehead three times and went back to his post at the door. I drank my coffee and then walked the floor back and forth. While the noise of the bells and guns continued, the wounded Mexicans lay as still as mice. I had no trouble with them at all. This was the only time that I ever had observed them all quiet at one time while I was with them.

After dusk Hale, Shirlock and Pete all came in. I asked them where they had been, they said they had been to see the old major. This was a Mexican officer who could talk good English, who appeared to be tolerably friendly to us, but still did not let us know but what their tales about the recapture of Santa Anna and the taking of Houston were facts. We were not long in learning that it was all done for fear we would rise and take the fort, which could have been easily done if we had known the facts.

The wounded were hurried off in carts and sent down to Copino, and placed on board of a vessel. When we were about to leave Goliad, the interpreter told us that Hale, Shirlock, Pete and myself could have our choice to go on with the wounded and their guard and remain loose, or we might be tied two and two together and walked to Matamoros with Miller's men. We four agreed to by sea.

"I appeal to those of my fellow-citizens who have exercised the profession of magistrate. They shall say how many times their trembling hands have signed a death sentence, the letters of which were blurred by their tears.."

—Antonio Lopez de Santa Anna in *Manifesto*, 1837

Isaac Hamilton, Survivalist

[Isaac D. Hamilton, an Alabama native, was one tough man. In 1852, he filed a petition with the state in an attempt to recoup the property and money that was taken from him during the Revolution. Appended to his petition were medical affidavits attesting to his wounds and an affidavit prepared by Hamilton giving the details of his ordeal. Not only did he survive the massacre, but he managed to survive half a month in the wilderness with two badly injured legs and yet another capture by the Mexicans. The affidavit attached to Hamilton's 1852 petition is given below. He further appealed to and corresponded with Representative James W. Henderson in early 1858 and again told his story. From this correspondence arose a statement from Dr. Joseph Barnard attesting to the fact that Hamilton was wounded at the Goliad massacre and escaped. He was eventually awarded a league of land near Beaumont in 1858, presumably to compensate for his losses, but he died in 1859 before the land was surveyed.]

Reminiscences of Goliad

During the spring of eighteen hundred and thirty five—May, I think—I became a member of Captain Jack Shackelford's Company then being organized in Alabama for the purpose of visiting Texas in her struggle for liberty, and in the fall of the same year we arrived at Passo Cavallo from which place we proceeded to Goliad where, after remaining some two months, we were ordered by Col. Fannin to retreat. We left Goliad and after marching some eight miles and within one and a half miles of the Coleto, were attacked by some fifteen hundred Mexicans under the command of Gen. Urrea in March 1836 and, there being but three hundred and sixteen of us in number, were overpowered by the enemy and compelled to surrender which we did, under the stipulations that we should be treated as prisoners of war and returned to the United States in safety, both in person and property.

With this understanding we were marched back to Goliad where on the sixth day after the battle, before sunrise we were marched some hundred yards from the fort. The first intimation we received that the articles of the treaty would not be complied with was an order—now in Spanish—for us to be shot. The question was asked if any one of us could speak Spanish and when answered in the negative, the order to "face about" was given in English—an order some obeyed while others refused. The order to fire was then given, the Mexicans being three paces distant.

I was shot in the left thigh about six inches above the knee. I then wheeled about and, while in the act of crossing the fence, received a bayonet in the upper portion of the right thigh after which I was pursued some three miles being frequently fired on and during which time I saw but one mounted Mexican who was endeavoring to rope one of our men. After having outrun my enemies I was hailed by a man by the name of Brooks and after having proceeded some four miles up the San Antonio River we were overtaken by two others—Cooper and Simpson—by whose assistance I was enabled to make my way on to within two miles of Texana, at which place the Mexicans then were, but being exhausted by the loss of blood and unable to proceed farther my comrades left me under the impression that surely I must die.

At this place I remained three days and nights, having nothing or little else on which to subsist save the dried matter on pus that run from my wounds, after which I crawled to Texana and, getting into a canoe, I proceeded to Dimmit's Point where on the nineteenth day after the massacre I was retaken by the Mexicans—who on the day previous had shot some twenty or thirty Indians and four or five white men. From this place I was hauled on a cart some fifteen miles then put upon a poor horse who, with myself, received alternately the Mexican quirt for the amusement of my tormenters until we arrived at Victoria. At this place I was court-martialed and ordered to be shot, which fate I escaped by the intercession of two Mexican Ladies.

I was then compelled to haul water from the river in the capacity of a horse and the reason I was not shot, as I understood, was that they wished to take me back to Goliad and have me shot there in accordance with the original order of Santa Ana. However, be that as it may, before I was well enough to march back, the news of the Battle of San Jacinto arrived & I and one other made our escape under cover of the night—being told by the Mexicans that we would be killed if for no other purpose than that of amusement.

Thus after making my way through many difficulties, dangers and privations I arrived at home in Alabama about the first of July 1836, having been about some fourteen months or rather in the service of Texas.

My losses sustained on the above occasion were, as nearly or I can now recollect, as follows, viz:

Twenty three hundred Dollars in actual cash $2300
Two horses and rigging one hundred dollars each $200
One trunk of clothing $175
Weapons of war and other valuable articles &c amounting in all to about three hundred one eighty dollars to the best of my recollections $380
$3,055.

In addition to the above I acted as quartermaster while under the command of Capt. J. Shackelford for which service I have never received any remuneration save the land to which I was entitled as a citizen of the Republic, nor have I ever applied for or received any part of the above amount of money and property lost during the invasion of Texas by the Mexican forces during the years eighteen hundred and thirty five and six.

—I. D. Hamilton.

Sworn to and subscribed before me at my office in Houston January 8th,1852. Witness my hand and seal of Harris County Court at my office.

—W. R. Baker, Clerk, Harris County.

Six Years Later, To Representative J. W. Henderson

Galveston
January 28th, 1858
Hon. J. W. Henderson

Yours of 20th came to hand this morning; contents duly noted. I belonged to Jack Shackelford's company called Red Rovers from Courtland, Alabama. Came a volunteer as quartermaster for our company, arrived in Texas in thirty-five, belonged to the service until the spring of thirty-six. Was with Col. Fannin at the Battle of Coleto and was marched back to Goliad; was shot thru the inner part of the left thigh & bayoneted thru the inner part of the right thigh at Col. Fannin's or the Massacre at Goliad after which I escaped by partly being helped over the brush fence by the point of a bayonet thru my right thigh…I was out nineteen days with out anything to subsist upon save what I could find.

I was in company with three of my companions the first eight days and walked between two of them with an arm round each of their necks to help me on my left leg. Was almost helpless. We were in about two miles of Texana when I gave out and was so weak I would become blind when they would attempt to raise me up, so they left me in that condition the eighth day after the Massacre.

I lay there or near the spot my companions left me for three days barely alive. Ate the matter from my wounds and tried to suck subsistence from green flies and lice of which I had a heavy stock. I began to think of the wolves and buzzards coming to prey up on my body before life was quite gone and I determined to make a start in the direction of an encampment where I last heard the sound of the Mexican trumpet which I gained by crawling and walking as I best could stand it.

That encampment was Texana. I lay hid round about there, five days, found some little substance to live upon. I made another attempt to get to some place where I might get some help but the effort was in vain. I only traveled some half mile and lay down under a live oak until the next day. I then returned to Texana.

After hiding about there one day longer, I raised a canoe in the stream and went to Dimmit's Point and was there taken prisoner by Don Placidore of Victoria. I was placed in the hands of some lancers and placed on a bare-backed horse and was most cruelly beaten until we arrived at Victoria. It would take too much space to speak of all that happened to me. I was sentenced to be shot at Victoria; two officers wives pled for me and I was to be reserved until I was able to be marched to Goliad, shot and burned according to Satana's order. I was compelled to draw water from the river in a barrel when I could scarcely walk. Was frequently straddled while I was lying down on my back by a Mexican with a knife in his hand drawing it across my throat with the oaths and motions as if my throat would be cut in a moment.

There is no words to express my sufferings. I was out nineteen days after I was shot and bayoneted at Goliad until I was taken by Placidore at Dimmit's point.

I lost my money and property by our defeat at Goliad, as near as I can recollect, five thousand dollars or thereabout. This looks like a hard storey but it is only a beginning to its reality.

You say Dr. Barnard will do all he can in conjunction with yourself for which I will be under obligation to you both and all who may assist in the matter. If I knew Dr. Barnard's address I would write him. Your attention to this matter shall never be forgotten by your old friend, humble servant,

Respectfully,
—I. D. Hamilton

Sworn to and subscribed before me at the City of Galveston this 28th day of January A.D. 1858. Witness my hand and official seal.

—James Welsh, Notary Public, Galveston County

"It has since been asserted that those at Goliad were executed with cruelty...I cannot be held responsible for the manner in which that officer carried out the law."

—Antonio Lopez de Santa Anna in *Manifesto,* 1837

Meet Dillard Cooper

[In the summer of 1870, Fannie A. D. Darden, a Texas poet and belle of Colorado County, interviewed Dillard Cooper about his escape from the Massacre. The account would be published in the Colorado Citizen *four years later. Cooper lived in the vicinity of Columbus at the time and Ms. Darden gives us a face to put with the voice we're about to hear, describing him thus: "He is a fine looking man, and although somewhat advanced in years, his appearance does not indicate that 'old father time' has dealt very hardly with him. He is over six feet in height, with light hair and blue eyes, and his form and proportions seem well fitted for the part he had to play in making his escape from the fiendish instruments of Santa Anna's cruelty." In addition to Isaac Hamilton, whose account appears in this text, Dillard Cooper fled in company with Wilson Simpson and Zachariah Brooks, by whom no known written accounts exist, other than the official statement referred to below, as published in a New York newspaper. Simpson took an active role in the service of the Republic after Goliad, participating in Indian raids in the San Saba country, as well as the Vasquez and Woll campaigns. Mr. Brooks left less of a paper trail and does not appear to have continued in the service much after Goliad. Cooper, Simpson and Brooks appeared before a notary in New Orleans on April 27, 1836, to attest to what they'd seen a month prior. Their statement appears in the* Morning Courier & New York Enquirer *of May 14, 1836.]*

On the morning of the 27th of March, 1836, about daylight, we were awakened by the guards, and marched out in front of the fort, where we were counted and divided into three different detachments, We had been given to understand that we were to be marched to Copano, and from there shipped to New Orleans. The impression, however, had in some way been circulated among us, that we were to be sent out that morning to hunt cattle; though I thought at the time that it could not be so, as it was but a poor way, to hunt cattle on foot.

Massacre

Our detachment was marched out in double file, each prisoner being guarded by two soldiers, until within about half a mile southwest of the fort, we arrived at a brush fence, built by the Mexicans. We were then placed in single file, and were halfway between the guard and the fence, eight feet each way. We were then halted, when the commanding officer came up to the head of the line, and asked if there were any of us who understood Spanish. By this time, there began to dawn upon the minds of us, the truth, that we were to be butchered, and that, I suppose, was the reason that none answered. He then ordered us to turn our backs to the guards. When the order was given not one moved, and then the officer, stepping up to the man at the head of the column, took him by the shoulders and turned him around.

By this time, despair had seized upon our poor boys, and several of them cried out for mercy. I remember one, a young man, who had been noted for his piety, but who had afterwards become somewhat demoralized by bad company, falling on his knees, crying aloud to God for mercy, and forgiveness. Others, attempted to plead with their inhuman captors, but their pleadings were in vain, for on their faces no gleam of piety was seen for the defenseless men who stood before them. On my right hand, stood Wilson Simpson, and on my left, Robert Fenner. In the midst of the panic of terror which seized our men, and while some of them were rending the air with their cries of agonized despair, Fenner called out to them, saying: "Don't take on so, boys; if we have to die, let's die like brave men."

At that moment, I glanced over my shoulder and saw the flash of a musket; I instantly threw myself forward on the ground, resting on my hands. Robert Fenner must have been instantly killed, for he fell with such force upon me as almost to throw me over as I attempted to rise, which detained me a few moments in my flight, so that Simpson, my companion on the right, got the start of me. As we ran towards an opening in the brush fence, which was almost in front of us, Simpson got through first, and I was immediately after him. I wore, at that time, a small, round cloak, which was fastened with a clasp at the throat. As I ran through the opening, an officer charged upon me, and ran his sword through my cloak, which would have held me, but I caught the clasp with both hands, and tore it apart, and the cloak fell from me. There was an open prairie, about two miles wide, through which I would have to run before I could reach the nearest timber, which was a little southwest of the place from where we started.

I gained on my pursuers, but saw, between me and the timber, three others, who were after Simpson. As I neared the timber, I commenced walking, in order to recover my strength, before I came near them. When he first started, we were all near together, but as Simpson took a direct course across the prairie. I, in order to avoid his pursuers, took a circuitous course. There were two points of timber projecting into the prairie, one of which was nearer to me than the other. I was making for the furthest point, but as Simpson entered the timber, his pursuers halted, and then ran across and cut me off. I then started for the point into which Simpson had entered, but they turned and cut me off from that. I then stopped running and commenced walking slowly between them and the other point. They, no doubt, thinking I was about to surrender myself, stopped, and I continued to walk within about sixty yards of them, when I suddenly wheeled and ran into the point for which I had first started. They did not attempt to follow me, but just as I was about to enter the timber, they fired, the bullets whistling over my head caused me to draw my head down as I ran.

As soon as I entered the timber, I saw Simpson waiting and beckoning to me. I went towards him, and we ran together for about two miles, when we reached the river. We then stopped and consulted as to the best way of concealing ourselves. I proposed climbing a tree, but he objected, saying that should the Mexicans discover us, we would have no way of making our escape. Before we arrived at any conclusion, we heard someone coming, which frightened us so, that I jumped into the river. Simpson ran a short distance up it, but seeing me, he also jumped in. The noise proceeded from the bank immediately above the spot where Simpson was, and I could see the place very plainly, and soon discovered that two of our companions had made their escape to this place. They were Zachariah Brooks, and Isaac Hamilton. Brooks was wounded in the hip. In the fleshy part of both Hamilton's thighs were wounds, one made by a gunshot and another by a bayonet.

We all four swam the river, and traveling up it a short distance, arrived at a bluff bank, near which was a thick screen of bushes, where we concealed ourselves. The place was about five miles above the fort. We did not dare proceed further that day, as the Mexicans were still searching for us, and Hamilton's wounds had become so painful as to prevent his walking, which obliged us to carry him. We remained there until about 10 o'clock that night, when we started forth, Simpson and myself carrying Hamilton. Brooks, though severely wounded, was yet able to travel. We had to proceed very cautiously and rather slowly.

Fort La Bahia being southeast of us, and the point we were making for, was about where Goliad now stands. We proceeded, in a circuitous route in a northeasterly direction. We approached within a short distance of the fort, and could not at first account for the numerous fires we saw blazing. We were not long in doubt, for the sickening smell that was borne towards us by the south wind, informed us too well that they were burning the bodies of our companions. And, here, I will state what Mrs. Cash, who was kept a prisoner, stated afterwards: that some of our men were thrown into the flames and burned alive. We passed the fort safely, and reached a spring, where we rested from our journey and from whence we proceeded on our travels.

But the night was foggy, and becoming bewildered, it was not long before we found ourselves at the spring from which we started. We again started out, and again found ourselves at the same place; but we had too much at stake to sink into despondency. So once more took our wounded companion, thinking we could not miss the right direction this time; but, at last when day began to break, to our great consternation, we found we had been traveling around the same spot, and were for the third time back at the identical spring from which we had at first set forth. It was now impossible to proceed further that day, as we dared not travel during the day, knowing we should be discovered by the Mexicans. We therefore concealed ourselves by the side of a slight elevation, amidst a thick undergrowth of bushes.

By this time, we began to grow very hungry, and I remembered an elm bush that grew at the entrance of the timber where we were concealed, which formed an excellent commissary for us, and from the branches of which we partook, until nearly every limb was entirely stripped. About 9 o'clock that morning, we heard the heavy tramp of the Mexican army on the march, and they not long after that passed within a stone's throw of our place of concealment. It seems indeed, that we were guided by an over-ruling providence in not being able to proceed further that night, for as we were not expecting the Mexican army so soon, we would probably have been overtaken and discovered by them, perhaps in some prairie, where we could not have escaped.

We remained in our hiding place the rest of the day, and resumed our journey after dark, still carrying our wounded companion. Whenever the enemy passed us, we had to conceal ourselves; and we laid several days in ponds of mud and water, with nothing but our heads exposed to view. When in the vicinity of Lavacca, we again got ahead of the Mexicans; and, after traveling all night, we discovered, very early in the morning of the ninth day, a house within a few hundred yards of the river. We approached it, and found the inhabitants had fled. When we entered the house, we discovered a quantity of corn, some chickens, and a good many eggs lying about in different places. Our stomachs were weak and revolted at the idea of eating them raw, so we looked about for some means of striking a fire, first searching for a rock, but failing to find one, we took an old chisel and ground it on a grindstone for about two hours, but could never succeed in getting the sparks to catch. We then concluded to return and try the eggs raw.

We had taken one, and Simpson was putting on his shoes, which he had taken off to rest his feet, which were raw and bleeding, and had just got one on when he remarked: "Boys, we would be in a tight place if the Mexicans were to come upon us now." So saying, he walked to the window, when to his horror, there was the whole Mexican army not more than a mile and a half off, and fifteen or twenty horsemen coming at full speed within a hundred yards of us. We took up our wounded man and ran to the timber, which was not far off, Simpson leaving his shoe behind him. We got into the timber and concealed ourselves between the logs of two trees, the tops of which having fallen together and being very thickly covered with leaves and moss, formed an almost impenetrable screen above and around us. We had scarcely hidden ourselves from view, when the Mexicans came swarming around us, shouting and hallooing through the woods, but did not find us. We heard them from time to time, all throughout the day and next night. The next morning, just before day, the noise of the Mexicans ceased, and we concluded they had left. Simpson then asked me to go with him to get his shoe, as it would be difficult for him to travel without it, and I consented to do so. We went out to the edge of the timber and stopped some time to take observations before proceeding further. Seeing nothing of the Mexicans, we proceeded to the house, found the shoe and, possessing ourselves of a couple of ears of corn and a bottle of water, we returned to our companions. We had no doubt that the Mexicans had gone, so we sat down and drank the water and ate an ear of corn, when Brooks asked Simpson to go with him to the house, saying he would get a chicken, and we could eat it raw. They started, and had hardly got to the edge of the timber when I heard the sound of horses feet, and directly afterwards the Mexicans were to be seen in every direction. I was sure they had captured Simpson and Brooks. Soon I heard something in the brush near us, but did not know whether it was the boys or Mexicans, but it turned out to be the boys, who crept undercover, and, in a few minutes, four Mexicans came riding by, passing within a few feet of where we were lying, with our faces to the ground.

After going into the woods a short distance they turned and passed out again, but it was not long after when six of them came riding quite close, three on each side of us, and leaning down and peering into our hiding place. It seemed to me they could have heard us, for my own heart seemed to raise me almost from the ground by its throbbings. I felt more frightened than I ever had been before; for at the time of the massacre, everything had come on me so suddenly that my nerves had no time to become unstrung as they now were. The Mexicans passed and repassed us, through the day, so we dared not move from our hiding place. A guard was placed around us the following night, the main body having, no doubt, gone on, and left a detachment to search for us. I think they must have had some idea of our being some of Fannin's men, or they would scarcely have gone to that trouble. About 10 o'clock that night we held a consultation, and I told my companions it would not do to remain there any longer, as the Mexicans were aware of our place of concealment, and would surely discover us the next day. We all decided then to leave, and they requested me to lead the way out. I told them we would have to crawl through the timber and a short piece of prairie, until we crossed the road near which the Mexicans were posted; that they must be careful to remove every leaf and stick in the path, and to hold their feet up, only crawling on their hands and knees, as the least noise would betray us to the enemy.

I was somewhat acquainted with the locality; for we were now not far from Texana, and I had sometimes hunted along these woods. Thus I led the way. Hamilton's wounds were so painful that we could move only slowly, and we must have been two hours crawling about 200 yards. When we at length passed the timber and reached the road, I stopped to make a careful survey of the situation. I could see the Mexicans placed along the road, about a hundred yards on each side of us. The moon was shining, but had sunk towards the west, which threw the shadow of a point of timber across the road, and concealed us from view. It would have been hard to discover us from the color of our clothes, as the earthy element with which they were mixed had entirely hidden the original fabric. We continued to crawl, until we reached a sufficient distance not to be discovered, when we rose up and walked. Although Hamilton had, with a great deal of pain, managed to crawl, yet it was impossible for him to walk, and his wounds had by this time become so much irritated and inflamed that he could scarcely bear to be carried. We traveled that night only a short distance, and hid ourselves in a thicket near a pond of water. Brooks had been trying to persuade me to leave Hamilton; but, although our progress was impeded by having to carry him, I could not entertain the idea for a moment. I indignantly refused, but still he would seize every opportunity to urge it upon me. He said it would be impossible for us to escape, burdened as we were with Hamilton. I could only acknowledge the truth of this, for it was a desperate case with us. The foe was around us in every direction, the main body being encamped at Texana, some two or three miles distant. Brooks, finding that I was not to be persuaded, then attempted to influence Simpson.

On the tenth day out, they took the bottle and went to the pond nearby, for water. As they were returning, (I suppose Brooks did not know he was so near the place they left us), both Hamilton and myself heard Brooks urging Simpson to leave him. He told him if we remained with Hamilton, we would certainly lose our lives; but there was some slight chance of escaping, if we left him, and that Hamilton's wounds had become so much worse that he was bound to die, unless he could have rest; and, as we were doing him no good, and ourselves a great deal of injury by carrying him, it was, our duty to leave him. Now Brooks had never carried Hamilton a step. Simpson and myself had done that; yet Brooks was the first to propose leaving him; and, although there was a great deal of truth in what he was saying, yet I felt quite angry with him, as I heard him trying to persuade Simpson. Hamilton did not say a word to them when they came in, but sat with his face buried in his hands a long time.

At length, he looked up, and said: "Boys, Brooks has told you the truth; I cannot travel any further, and if you stay with me, all will be killed. Go and leave me, boys; if I have rest I may recover, and if I ever should get off safe, you shall hear from me again." He spoke so reasonably, and we were so thoroughly convinced of the truth of what he said, after a brief consultation, we decided to depart without him. Hamilton had known Brooks in Alabama; he called him over, and gave him a gold watch and $140 in gold, telling him to give it to his mother. We then bade Hamilton farewell, all of us shedding tears as we parted, but when we turned to go, my resolution failed me, and I could not find it in my heart to leave him. I said: "Boys, don't let us leave him." But Simpson and Brooks said that we could do neither him nor ourselves any good by remaining, and that they were determined to go. I told them I would remain with him, and do the best I could for him. So they started off without me; but Hamilton insisted so much that I should leave him, that I again bade him farewell, and followed and soon overtook the others. The reason that we started off in the day was that it was raining quite hard, and we thought there would not be much danger in traveling, but we had not gone more than halfway through the next prairie, when the weather cleared up, and we saw the whole Mexican army encamped at Texana, about two miles off. But they did not discover us, and we succeeded in reaching the timber on the Navidad. In the evening we walked out to a slight eminence which overlooked the prairie, to reconnoitre. While gazing across the prairie, we could see three men on horseback, but so indistinct were they, that we could not at first tell whether they were Americans or Mexicans. As they approached, we hid in the undergrowth, and as they passed, we saw that they were Mexican couriers returning to the command.

At eight we again started forth, and coming out on the prairie, we discovered a road, which we concluded had been made by the refugees in their retreat from the enemy. During all this time we had nothing to eat but leaves and herbs, and the two ears of corn that we got at the house on Lavacca river. On the twelfth day, we reached the Colorado, at Mercer's crossing. As we were very tired, we sat down on the bank to rest a little, before attempting to swim over. While sitting there, a dog on the opposite side of the river began to bark. When we heard that well-known sound, our very souls thrilled with joy, and that was the first time since the awful day of the massacre that a smile had ever illuminated our faces. We looked at each other and then burst into a great big laugh. We were all good swimmers, but I sometimes took the cramp while swimming, so we concluded to cross on a log. We procured a dead mulberry pole, and hanging on to it, one at each end, and one in the middle, we crossed over.

We went up to Mercer's sugar house, which was not far off, where we found a number of hogsheads of sugar and molasses. As soon as we discovered the sugar, we commenced eating it, but in a minute or two we were taken very sick and threw it up. We tried it again, with like success, so we concluded that sugar was not the thing for starving men, and started for the house, which was about a mile distant. Not a soul was to be seen about the place. We entered the house and saw everything just as the inhabitants had left—the furniture in its place, and the table neatly set for two persons. To our great surprise, there was a fire burning in the fireplace, and as we approached, we saw an oven and skillet sitting on the hearth. I lifted the lid off the oven, and saw there three loaves of bread. I looked into the skillet, and beheld a chicken, nicely prepared and still warm. It looked like the magic of some fairy tale. We did not stop to ask questions, but proceeded to help ourselves. But neither the chicken or the bread agreed with us any better than the sugar, though we only eat a very small portion of them. I then boiled some eggs about half done, and as we only ate a half a one apiece, we succeeded in retaining it.

While we were eating our eggs we saw approaching us a man and woman on horseback, and as they drew near, we perceived them to be a mulatto man and one of the most beautiful white women I ever saw. He informed me that they had been there and prepared the meal of which we had partaken, and had rode off to look for his wagon, which had been lost in the river. The woman had nothing to say, but looked very much cowed. The Negro watched her very closely when in our presence. But at length he had to go and attend to his horses, so that she was left alone with us. We expected then that she would speak, and perhaps appeal to us for protection, and we had agreed that if she did, we would afford it; but she said nothing to us, and we did not like to question her.

That night before we retired to rest, we agreed that one of us should stay awake and watch while the others slept. We had not slept long when the watcher woke us, saying the Negro had left the house. We immediately arose, and following him, demanded of him the cause of his leaving. He said he had forgotten to water the horses, and had started out for that purpose; but we knew he had watered them, and therefore ordered him to return, threatening to kill him if he refused. We ought to have killed him then, for we afterwards learned that he had stolen the poor girl, as she was flying with her family from the Mexicans, and she said afterwards that she wished to appeal to us for protection, but feared that we would refuse it, and he would murder her, as he had threatened to do. He afterwards joined Santa Anna and was killed at the battle of San Jacinto.

We traveled the succeeding day and night, resting but little, and reached the bank of the Brazos on the morning of the fourteenth day; and on the west side of the river we met three white men carrying dispatches to Houston's army. One of them having lamed his horse, turned and went down the country with us to within about twelve miles of Columbia to a house where we feasted like princes. As I have had nothing but hard fare to regale my readers with, I cannot refrain from giving our bill of fare on this occasion. It consisted of chickens, eggs, bacon, corn and flour bread, brandy and whisky, and cheese for dessert. After recruiting a little, we procured horses, with the intention of joining Houston's army; but before we reached there the Battle of San Jacinto had been fought and won.

It was more than a year before I ever heard anything of Hamilton. He remained in the same place where we left him nine days, sometimes lying in the pond of water, which assuaged the pain of his wounds. At the end of that time he was so much improved that he essayed to walk to Texana, which the Mexicans had evacuated, and succeeded in doing so. When he reached there, all was still and deserted. While standing near a warehouse, he saw two Mexican soldiers approaching, and being too weak for flight, he slipped into a warehouse and got behind a dry good's box, which not being a very large one, left his head and shoulders exposed to view. The Mexicans soon came up, and while sitting on their horses, looked in and fixed their eyes directly on him. He sat perfectly motionless, not moving a muscle, and the Mexicans, after looking at him awhile, burst into a loud laugh and rode off.

He left there that evening and spent the night under a live-oak, thinking that he would try and make his way through the country; but the next morning changed his intention, and taking a skiff made his way down to Dimmit's Landing. He said the best eating he ever had in his life was when he first entered Texana, and ate the meat from the rawhides the Mexicans had left. He had scarcely reached Dimmit's Landing, when he was perceived and taken prisoner by a Mexican soldier. Not long after other soldiers came in and, tying Hamilton on a mule, started for camp. He suffered so much from his wounds that he fainted several times on the way. Whenever this occurred, they would untie him, lay him on the ground, and throw water in his face until he revived, when they would again mount him on the mule, and proceed on their way. When they arrived at camp, he found two other Texians who had been captured. Hamilton remained in their hands for some time, and gradually grew well of his wounds.

There was a Mexican who waited upon him, who seemed to grow much attached to him, and Hamilton was led to place confidence in him. One evening this Mexican told him that if he wished to live another day, he must attempt to make his escape that night, as he had learned that he and the other prisoner were to be shot before morning; but at night the Mexican came to him again, and informed him that news had been received from the army which postponed the execution. In a day or two he came again with the information that they were to be shot very soon. Hamilton then arranged a plan for escape for himself and two companions.

The next evening, as had been agreed on, Hamilton walked towards the timber a short distance off and commenced gathering moss from the trees—one of his companions following him about fifty yards off, and the other fifty yards behind him, but by some means the last man disappeared, and Hamilton never knew what had become of him. The two commenced gathering moss, and when they had procured a load apiece, they began very slowly to return. It was now nearly dark, and when they had returned a short distance, they threw down the moss and ran; and as they did so, they looked back and saw some Mexicans soldiers pursuing them. There were some Mexicans in front of them driving a caballado of horses. Hamilton and his companion ran into this drove, causing them to stampede. One gentle horse stood still. They both mounted him and urging him along with their feet, and guiding his course by striking him with their hands on the side of the head, they soon outstripped their pursuers, who were on foot. They continued to ride until the horse fell beneath them. The next day they were fortunate enough to secure another, and after manufacturing a grapevine bridle and moss saddle, continued their journey, alternately walking and riding, and in this way finally succeeded in escaping.

Thus ends my account, though I design, at some future time, giving to the world other particulars relating to that eventful period, and the subsequent adventures of my companions and myself.

"Well, gentlemen, in eight days,
liberty and home!"

—Col. Juan Jose Holzinger

Joseph Barnard, Fighting Doctor

[Dr. Barnard's account appeared in several publications in the late 19th century and early 20th. Historian Hobart Huson even stitched them all together in 1949 to assemble the best possible composite of Barnard's published experiences. In the interest of choosing the publication closest to the actual events, we have selected the first published form, the serial publication of Barnard's narrative from the Goliad Advance-Guard *of 1875, as republished in Linn's* Reminiscences *in 1883.]*

Whenever he had an enterprise of activity that required prompt courage to overcome danger and win success, Fannin had sustained himself and acquired the confidence of his men. But his position was getting perplexed now and dangerous. The loss of Grant and Johnson's men, the now certain anticipation of the loss of the Alamo, and the delay of the people turning out to defend the frontier, where so much was at stake, served to cover our prospects with a gloom that deepened more and more, and was destined not to be expelled until after unparalleled scenes of barbarity and murder had been perpetrated and the country deluged with blood.

Thursday, March 10, 1836, a party of sixteen or eighteen of Captain Shackelford's company, under the command of Lt. Francis, were ordered down to the ranchos occupied by the citizens of Goliad, and about fifteen miles below the town. The principal object was to investigate some reports of there being Mexican soldiers among them as spies, who were gaining intelligence of our movements and designs, to report the same to the enemy. Tired of the monotony of life in the fort, and wishing for some exercise in the open country, I obtained leave to go with them. The expedition was of no importance in itself, and would have passed into oblivion had it not been that I preserved the notes I then made, which thus enabled me with certainty to fix the dates of subsequent occurrences. I had kept a regular journal up to this time, together with all other papers and some clothing, at the time of our battle and capture. These notes had remained in my pocket where I providentially discovered them after losing everything else, and have carefully preserved them until this time…

March 12th—News had also come in of the fall of the Alamo and the slaughter of every one of its defenders. About this time, certainly before today, arrived the order from General Houston to Colonel Fannin to retreat to Victoria. This was the first and only communication had from General Houston while Fannin was at Goliad; in fact, it was the first information we had of his whereabouts. The necessity of a retreat was now palpable to all. So far from Colonel Fannin wishing to disobey the order, I know this from his own lips that he intended to conform to it as soon as the Georgia battalion should return; and I had heard him before this express a wish that General Houston would come on and take command of the troops. The alleged disobedience of Colonel Fannin to Houston's order is an undeserved censure on a gallant soldier; and that he wrote back a refusal I know to be false. Circumstances have enabled me to possess a positive knowledge on these points and justice to both the dead and the living requires of me thus to state it.

…Thursday, 17th—We were now in a state of intense anxiety respecting the fate of our comrades. Nothing had been heard from them since they left us on Saturday morning, and none of our messengers had returned. We were convinced that some calamity had befallen them, and of its nature and extent we now had gloomy apprehensions. At length, about 4 p.m., Captain Frazer, true to his word, arrived and gave us full and explicit information and the following purport:

Colonel Ward had reached Refugio and relieved Captain King. Instead of immediately turning back, they unfortunately delayed their return, and Captain King started off with his company to destroy some ranches where the people had shown some hostility.

Colonel Ward was soon after attacked by some Mexican troops and driven into the church from which he had but a short time before released Captain King. He now found that Urrea, with his whole division, was about him and endeavoring to dislodge him from his position in the church. Ward and his men gallantly defended themselves, and repulsed all attacks made upon them. When night came, finding their ammunition exhausted, they succeeding in eluding the vigilance of the Mexicans and, silently leaving the church, retreated to the coast.

Captain King, upon reaching the ranches that he intended to destroy, met with opposition and got rather worsted in the fight. He made a circuit to get back to Refugio, which he reached in the night, and found the town occupied by the Mexicans. They then crossed the river and endeavored to retreat from the place but got lost on the prairie and, after wandering all night, found themselves in the morning at a place called Malone's Ranch. They had been watched and followed by a party of spies, and soon a force was around them that made resistance hopeless.

They surrendered and were immediately started in the direction of Goliad. They had proceeded but a few hundred yards when a halt was made in the prairie. King and his comrades were taken out and shot. Such were the results of this expedition.

Fannin and his officers immediately held a council and, without any hesitation, resolved to commence our retreat the next morning. Hardly had they left the council room when some of our scouts came in with the information of a large force of the enemy in the vicinity. Preparations were made for leaving and patrols were kept out all night to watch and give notice of any movement that might be made by the enemy.

Friday, 18th—This morning, while taking the necessary measures for a retreat in accordance with the officers in council last evening, a party of the enemy were discovered reconnoitring in the vicinity of the fort. Colonel Horton and a few horsemen sallied out to engage them. They did not wait for an attack, but fled, followed by Horton, until a large body of the enemy appeared who, in turn, chased the Texans back. Horton then sent for all the horsemen in the fort, who turned out to his assistance and enabled him to resume the offensive.

As the affair was nearly all visible from the fort, it produced considerable excitement, and all left their work to witness the "sport." I went with several others to the top of the church, which commanded a fine view of the country for miles around. Colonel Horton, now giving chase to his late pursuers, followed them over to the north side of the river and on over the prairie; occasionally a shot was fired, until the parties were lost from our view in the distance.

After a time, they made their appearances coming back; but the condition of affairs had changed: the Mexicans were pursuing our men and pressing them rather hard. But they succeeded in reaching the old *Acanama*, a mission, and, getting under shelter of its walls, made a stand. This was one of the old Spanish missionary stations, now in ruins, and standing about three-fourths of a mile northerly from the fort of La Bahia, and on the north side of the river. The Mexicans, numbering about one hundred, drew up in front at a safe distance, and opened fire, which was returned. Captain Shackelford now started out with his company to relieve Horton, and our artillerymen got one of their guns mounted on the wall and brought to bear on the Mexican party. A shot was fired at them, which fell short; but they deemed it wise to withdraw. They soon disappeared and we saw no more of them.

Colonel Horton left his position to return, and met the company going to his relief with warm greetings (they had forded the river and gone about halfway to him), and with them returned to the fort in the highest spirits. The events of the day had animated all, and good humor and cheerfulness for a while prevailed. Thus far the events of the day were, perhaps, beneficial; but alas! other considerations were forced unwillingly on our minds. The day was spent, when time with us was precious, and our retreat was necessarily postponed until the morrow. The horses were tired and jaded down, and our oxen that had been gotten up to draw the cannon and baggage carts were left all day without food; and we had given the enemy a day of our invaluable time in which to select his positions and perfect his arrangements for our destruction. I never heard that any man had been injured on either side in the foregoing skirmishing.

Although fully determined, from the necessity of the case, on retreating, we were by no means disposed to run. We confidently counted on our ability to take ourselves and all our baggage, etc., to Victoria. We still had about two hundred and seventy men, besides Colonel Horton's company of about thirty horsemen, numbering in all about three hundred; and, though mishaps had come on thick and fast, we still had confidence and determination. The necessary guards were posted for the night, during which we had some rain. Some alarms occurred but they proved groundless.

Saturday, 19th—The morning opened with a heavy, impenetrable fog. We left the fort as early as possible, with all our artillery and baggage, which was drawn by oxen. We fully expected an attack at the ford of the river. Colonel Fannin had, however, despatched Colonel Horton at an early hour to go down and occupy such position as would most effectually prevent the enemy from interrupting us in the passage.

We succeeded in crossing the river without molestation, but with some delay arising from the weak condition of our teams. After all had passed, Colonel Horton's company was directed to scout around and give us notice should any of the enemy appear; for as yet, none of them had appeared and we were still favored by the fog, which was very heavy.

We then marched on six miles to the Manahuilla Creek. After passing that about a mile, we came upon a patch of green grass where the prairie had been recently burned, and we halted to allow the animals to graze, as well as for the men to partake of refreshments. Our scouts reported no enemy discoverable within four or five miles. No manifestations of an attack, nor even of a pursuit, were apparent. After an hour's halt, we resumed the march, supposing that now the enemy did not intend to obstruct our retreat, as they had neglected to avail themselves of the most suitable position for harassing us; and relying on the alertness and fidelity of our horsemen for giving us timely notice of their approach, we proceeded about two miles. Our teams showed signs of weariness and our march was necessarily slow.

We had reached a low ridge when we discovered the enemy advancing in our rear. They had just emerged from the belt of timber that skirted along on this side of the creek, and consisted of two companies of cavalry and one of infantry. We halted, and a six-pounder was unlimbered, from which three shots were fired at them, but, as we perceived, fell short. It appears that four horsemen had been left in the rear and that they, instead of keeping a lookout, had, under a false sense of security, lain down and were only aroused by the close approach of the Mexicans. They now came at full speed; one of them, and one only (a German of the name of Ehrenburg), joined us. The other three, in the greatest apparent terror, passed about a hundred yards on our right, without even stopping to look at us, and under the strongest appliances of whip and spur, followed by a few hearty curses from our men.

Observing one or two more bodies coming from the woods, Colonel Fannin ordered his men to resume the march slowly, so as not to harass the jaded oxen, saying further that they enemy in sight were merely the skirmishers, etc.; that Colonel Horton, being notified by our firing that the enemy were in sight, would immediately return and rejoin us, and that we had only to keep ourselves cool and collected, and we could easily foil such a party. The men all viewed the matter in the same light, and we marched on coolly and deliberately for about one mile further, expecting all the time to see our horsemen coming to join us. We had now come to a piece of low ground, and were yet about half a mile from the point of timber, when we were brought to a halt by the breaking down of our ammunition cart.

One company of the enemy's cavalry had come up abreast of us on our right flank, and the others had got a little in advance on the left, their infantry coming up in our rear. Before we could make any disposition of our broken cart, they closed around our front and opened fire, and in this way the battle commenced. Colonel Fannin directed the men to reserve their fire until the enemy was near enough to make sure shots. Soon, however, the fire became general on our side as well as theirs.

I judged the enemy to be about five hundred strong at the commencement, but other troops kept continually coming up during the engagement, and by night they had not less than one thousand men opposed to us. The enemy's cavalry made numerous attempts to charge us, forming behind a little rise in the ground about four or five hundred yards off, then galloping up at full speed. But they were always so warmly received by our rifles that they were obliged to fall back. So confident were we in the beginning of the affair that Colonel Horton and his men would come back and rejoin us, that in several of their charges a number of our men, imagining the party to be Horton's troop, called out, "Don't fire! They are our horsemen!"

But neither Horton nor his men ever made their appearance. Our artillery did not appear to have as much effect on the enemy as we expected, and after the brave Petreswich, who commanded it, fell, and several of the artillerymen wounded, the guns were not much used in the latter part of the fight. Our men behaved with the utmost coolness and self-possession; and when it is considered that they were undisciplined volunteers, and this the first time (in most cases) of their encountering an enemy, their order regularity would have reflected credit on veterans. The fight continued without intermission from about 3 p.m. until night caused a cessation. The enemy drew off to the timber and encamped, having us surrounded by numerous patrols.

We now had time to look around and consider our situation. It was sunset and a night of impenetrable darkness, such as is rarely witnessed, succeeded. We were without water, and many, especially the wounded, were suffering from thirst. Upon further inquiry we found, from some unaccountable oversight, we had left our provisions behind. Our teams, during the engagement, were killed or had strayed off beyond our reach. We had seven men killed and sixty wounded, about forty of whom were disabled. Colonel Fannin had committed a grievous error in suffering us to stop in the prairie at all. We ought to have moved on at all hazards and all costs until we reached the timber. We might have suffered some loss, but we could have moved on and kept them at bay as easily as we repulsed them while stationary.

Fannin behaved with a perfect coolness and self-possession throughout and evinced no lack of bravery. He was wounded in the thigh, and had the cock of his rifle carried away by a musket ball while in the act of firing. His former experience in fighting Mexicans had led him to entertain a great contempt for them as soldiers, and led him to neglect to take such precautionary measures as were requisite from their great numerical superiority. On leaving Goliad, I had taken my spare clothing and papers and rolled them up in my blanket, which I slung on my shoulders as a knapsack; and at the beginning of the action, finding that it somewhat embarrassed my motions, I took it off and threw it in the middle of the square. Now, on looking for it, it was gone, and I saw no more of it. On account of the excessive darkness and our having no lights, I found no chance to attend to the wants of our suffering men to any extent. The want of water was most severely felt, for all had become thirsty, and more especially the wounded, whose misery was greatly aggravated by it.

It was determined by the officers to wait until morning before any further action was attempted; indeed it was impossible to do otherwise under all the circumstances. In addition to the excessive darkness the air was misty, and not a breath of wind, and it would have been impossible to keep together or follow a straight course for two hundred yards.

Weary and supperless, I lay down on the bare earth, without any cover, in order to obtain some repose; but the coldness of the ground soon benumbed my limbs and roused me from an unsatisfactory slumber to seek warmth in some exercise. This was supplied by an order to make an entrenchment. During the fight, while drawn up in order of battle, which was a hollow square, we occupied about an acre of ground. When the firing ceased, we had left the lines and congregated in the center, where we lay down. The entrenchment was made around us as we then were, and did not enclose a fourth part of the ground we occupied in the battle.

We set to work with our spades and dug a ditch two or three feet in depth. Our carts were then drawn up and disposed of upon the breastworks, so as to aid in our protection; and the carcasses of two horses, all that we had with us, and those of several oxen, were piled up for breastworks. Thus the night wore away, the enemy's patrol keeping up incessant music with their bugles to regale us; while the shrill and discordant scream of *"Sentinel alerto!"* which afterwards became so familiar, then first jarred upon my ear.

I worked with the spade until fatigued, then lay down for a little troubled sleep until the chilly earth forced me to seek for warmth by using the spade again. In such alternations the dismal night wore away, and day at last dawn upon us. It was Sunday, March the 20th. Early in the morning, and before it was quite light, we perceived a reinforcement of three or four hundred men coming to the enemy, accompanied by a hundred pack mules. They brought up two pieces of artillery and a fresh supply of ammunition, and they directly commenced the business of the day by treating us to a few rounds of grape and canister. The enemy now being well-supplied, their force so superior to our own—having at least one thousand three hundred men in good order, while we, exclusive of our wounded, could only must about two hundred, and they worn out by the toils and exertions of the previous day—left our situation perilous in the extreme.

The question was now agitated: "Should we surrender?" We well knew their faithlessness and barbarity, as shown in the recent example of Johnson and King, and that we could not rely on any feelings of honor or humanity in them when once they had us in their power. The only chance for us to escape from them was by a desperate rush through their main body into the timber. This would necessarily involve the abandoning of our wounded to a certain death and leaving everything behind. We felt confident, indeed, in being able to keep them at bay; but without provisions or water, it would only be to postpone without averting our fate, and we were now satisfied that no aid would come to us from Victoria or the settlements.

The officers consulted together and then submitted the question to their respective companies. I was with my messmates in Shackelford's company when he submitted the proposition to us. After a cool discussion of the chance it was considered that if the enemy would agree to a formal capitulation, there would be some chances of their adhering to it, and thus saving our wounded men. Dr. Shackelford resolutely declared that he would not agree to any alternative course that involved an abandonment of his wounded men. It was finally agreed that we would surrender if an honorable capitulation would be granted, but not otherwise, preferring to fight it out to the last man in our ditches rather than place ourselves in the power of such faithless wretches without at least some assurance that our lives would be respected.

These, as understood, were the sentiments of the party generally. When the matter was first proposed to Colonel Fannin, he was for holding out longer, saying, "We whipped them off yesterday, and we can do so again today."

But the necessity of the measure soon became obvious. He inquired if the sentiment was unanimous, and founding that all, or nearly all, had made up their minds, he ordered a white flag to be hoisted. This was done, and was promptly answered by one from the enemy. The flags met midway between the forces. Colonel Fannin, attended by Major Wallace, the second in command, and Captain Dusanque, as interpreter, went out and met the Mexican commanders. After some parley, a capitulation with General Urrea was agreed upon, the terms of which were that we should lay down our arms and surrender ourselves as prisoners of war; that we should be treated as such according to the usage of civilized nations; that our wounded men should be taken back to Goliad and properly attended to, and that all private property should be respected.

These were the terms that Colonel Fannin distinctly told his men, on his return, had been agreed upon, and which was confirmed by Major Wallace and Captain Dusanque, the interpreter. I saw Colonel Fannin and his adjutant, Mr. Chadwick, get out his writing desk and paper and proceed to writing. Two or three Mexican officers came within our lines and were with Colonel Fannin and Chadwick until the writing was finished. We were told that the articles of capitulation were reduced to writing and signed by the commander of each side and one or two of their principal officers; that the writings were in duplicate, and each commander retained a copy.

I am thus particular and minute in regard to all the incidents of this capitulation, and especially what fell under my personal observation, because Santa Anna and Urrea both subsequently denied that any capitulation had been made, but that we surrendered at discretion. We were also told, though I cannot vouch for the authority, that as soon as possible we should be sent to New Orleans under parole not to serve anymore against Mexico during the war in Texas; but it seemed to be confirmed by an observation of the Mexican colonel, Holzinger, who came to superintend the receiving of our arms. As we delivered them up, he exclaimed, "Well, gentlemen, in ten days, liberty and home!" Alas! within that time, most of our men did reach their final homes.

We now surrendered our arms, artillery, ammunition, etc., to the Mexicans, who took immediate possession. Our officers were called to put theirs by themselves, which we did, in a box which was nailed up in our presence, with the assurance that they should be safely returned to us on our release, which they flattered us would shortly take place.

Now that our fate was decided, I gave all my attention to the wounded. I was assisted by Dr. J. E. Fields, who had joined us about ten days before; also by Dr. Shackelford, captain of the Red Rovers, who was a surgeon and physician by profession; and by Dr. Ferguson, a student of his, who had come out with his company. The prisoners were now marched back to Goliad, the wounded being left on the field until carts could be sent for them.

The loss of the enemy in the engagement I could never learn with precision. They had above a hundred wounded badly that we (the surgeons) were afterwards obliged to attend to. Fifteen of their dead were counted within a few hundred yards of our entrenchment early in the morning, besides an officer, badly wounded, who was brought into our camp and died shortly after. The accounts of the Mexicans themselves, of whom I subsequently inquired, varied in their statements of their dead from forty to four hundred. Thus terminated the Battle of Encinal del Perdido, by which, from untoward events, we were placed in their power; yet they had but little cause to boast of their victory.

Monday, March 21st—Carts came out and took in a portion of our wounded men, attended by the other surgeons, while I remained on the ground with those left. This day, while dressing the wounds of our men, some of the soldiers stole my pocket case of instruments, and thus deprived me of the power of properly attending them.

Tuesday, March 22nd—Carts came again today and took in the remainder of our wounded. Captains Dusanque, Frazer and Pettus, and two or three other men who had been left with me on the ground, went also with the last of the wounded.

At the Manahuilla Creek, we met General Urrea with about one thousand men going to Victoria. The captain of the escort appeared to be a very gentlemanly man, and endeavored to cheer up our spirits. Finding that Captain Dusanque could speak Spanish, he engaged him in lively and cheerful conversation, dismounting and walking with us for several miles. We certainly were inspired with more confidence by his lively and cheerful manners. It was dark when we reached the San Antonio, which we waded, it being three feet deep. Perceiving some disorder among the carts which had not yet crossed, our Mexican captain went back to them, and the guard halted a moment on the east side. Captain Dusanque now remarked in a very serious tone that contrasted strangely with the cheerful voice in which he had been conversing: "I am now prepared for any fate." The words and his manner struck us with surprise, and he was asked if he had ascertained by anything the captain had said that treachery was meditated.

"No," he replied, and ominously repeated his former remark. The idea struck me that here was a chance to escape by silently dropping into the water while the guard and their captain were on the other side, and from the darkness could not see me; in two or three minutes I would have floated beyond their reach, and, being a good swimmer, could then easily escape. I stopped to consider the matter more fully, and directly the captain and his guard were alongside of us; and thus by indecision in a critical moment, I lost the chance.

After, events frequently called this matter to my mind and made me bitterly regret not having acted on the first impression of my mind. It was late when we reached the fort, and we were sent into the church, where we found all the prisoners were placed and crowded up in a very uncomfortable manner, strictly and strongly guarded.

Wednesday, 23rd—My first effort was to see Colonel Fannin, and if by any possibility through him get hold of some of our surgical instruments and hospital dressings for the wounded, we having been robbed of everything of the kind. Most of such articles had belonged to individuals; and Colonel Fannin, at my request, addressed a note to the Mexican commandant, in which he claimed sundry instruments and other articles, not only as private property according to the articles of the capitulation, but from the necessity of the surgeons having them for the benefit of the wounded Mexicans as well as of the Americans. The application was of no avail, and I should not mention it except to show that the terms of capitulation had been appealed to once by Colonel Fannin, which, of course, he never would have done had there been no capitulation. This day all the prisoners except the wounded were removed from the church and placed on the west side of the fort. The church being still too small, the American wounded were removed to the cuartels [quarters] on the west wall.

Thursday, March 24th—We had been politely requested by the Mexican officers as a favor that we would attend to their wounded, as their surgeons had not arrived, while we, not to be outdone by them in politeness, replied that we would do so with the greatest pleasure. We, however, found that we were not to be permitted to visit our own wounded until we had attended to all theirs. We remonstrated against this arrangement, but to no purpose.

A Mexican surgeon at length arrived, but we had no assistance from him. It took us nearly the whole day to get through with the Mexicans before we could be allowed to see our own men; and then we had so little time that we could only dress some of the severest wounds and leave the rest altogether. Some of them up to this time had not the first dressing. We resolved to refuse attendance altogether upon the Mexicans, at all risks, unless we could be allowed time enough to properly attend to our own men at least once a day. But at this time, Major Miller, with seventy men, who had come from Nashville, Tennessee, and who had been made prisoners on their landing at Copano, were brought in. Major Miller immediately tendered his services to us as a medical aid, as did some of his men, by which our labor was much lightened, and we thenceforward managed to get along without an open rupture with our taskmasters.

Saturday, March 26th—Colonel Fannin, who with his adjutant, Mr. Chadwick, had been sent to Copano, returned this day. They were placed in the small room of the church which had been appropriated to the surgeons and their assistants and guard—rather crowded to be sure, but we had become accustomed to that. They were in good spirits and endeavored to cheer us up. They spoke of the kindness with which they had been treated by the Mexican colonel, Holzinger, who went with them, and their hopes of our speedy release. Fannin asked me to dress his wound, and then talked of us his wife and children with much fondness until a late hour. I felt more cheerful this evening than I had before since our surrender. We had reiterated assurances of a speedy release, it is true, by the Mexicans, though we placed but little reliance on them.

Our fare had been of the hardest, being allowed no rations except a little beef or broth. Now we had been able to purchase from the camp followers a little coffee and bread, more grateful to me than any luxury I ever tasted; and, after sleeping on the ground without a blanket from the time of our last capture, I had at last succeeded in getting an old worn out one, upon which I had lain down to rest this evening with more pleasure and happier anticipations than I had before allowed myself to indulge in.

Sunday, March 27th—At daylight, Colonel Garay, a Mexican officer, came to our room and called up the doctors. Dr. Shackelford and myself immediately rose (Dr. Field was at a hospital outside the fort) and went with him to the gate of the fort, where we found Major Miller and his men. Colonel Garay, who spoke good English, here left us, directing us to go to his quarters (in a peach orchard three or four hundred yards from the fort) along with Miller's company, and there wait for him. He was very serious and grave in countenance, but we took little notice of it at that time. Supposing that we were called to visit some sick or wounded at his quarters, we followed on in the rear of Miller's men. On arriving at the place, Dr. Shackelford and myself were called inside a tent, where were two men lying on the ground, completely covered up, so that we could not see their faces, but supposed them to be the patients that we were called in to prescribe for. Directly a lad came in and addressed us in English. We chatted with him for some time. He told us his name was Martinez, and that he had been educated at Bardstown, Kentucky.

Beginning to grow a little impatient because Colonel Garay did not come, we expressed an intention of returning to the fort until he would come back; but Martinez said that the directions for us to wait there were positive, and that the colonel would soon be in, and requested us to be patient a little longer, which was, in fact, all that could be done. At length we were startled by a volley of firearms, which appeared to be in the direction of the fort. Shackelford inquired, "What's that?" Martinez replied that it was some of the soldiers discharging their muskets for the purpose of cleaning them.

My ears had, however, detected yells and shouts in the direction of the fort, which, although at some distance from us, I recognized as the voices of my countrymen. We started, and, turning my head in that direction, I saw through some partial openings in the trees several of the prisoners running at their utmost speed, and, directly after, some Mexican soldiers in pursuit of them.

Colonel Garay now returned, and, with the utmost distress depicted on his countenance, said to us, "Keep still, gentlemen; you are safe. This is not from my orders, nor do I execute them." He informed us that an order had arrived the preceding evening to shoot all the prisoners; but he had assumed the responsibility of saving the surgeons and about a dozen others, under the plea that they had been taken without arms. In the course of five or ten minutes, we heard as many as four distinct volleys fired in as many directions, and irregular firing which continued an hour or more before it ceased. Our situation and feelings at this time may be imagined, but it is not in the power of language to express them. The sound of every gun that rung in our ears told but too terribly of the fate of our brave companions, while their cries, which occasionally reached us, heightened the horrors of the scene.

Dr. Shackelford, who sat by my side, suffered perhaps the keenest anguish that the human heart can feel. His company of Red Rovers, that he had brought out and commanded, were composed of young men of the first families in his own neighborhood—his particular and esteemed friends; and besides two of his nephews, who had volunteered with him, his eldest son, a talented youth, the pride of his father, beloved of his company, was there; and all, save a trifling remnant, were involved in the bloody butchery.

It appears that the prisoners of war marched out of the fort in three different companies: one on the Bexar Road, one on the Corpus Road, and one towards the lower ford. They went one-half or three-fourths of a mile, guarded by soldiers on each side, when they were halted, and one of the files of guards passed through the ranks of the prisoners to the other side, and then all together fired upon them. It seems the prisoners were told different stories, such as they were to go for wood, to drive up beeves, to proceed to Copano, etc.; and so little suspicion had they of the fate awaiting them that it was not until the guns were at their breasts that they were aroused to a sense of their situation.

It was then—and I proudly record it—that many showed instances of the heroic spirit that had animated their breasts through life. Some called to their comrades to die like men, to meet death with Spartan firmness; and others, waving their hats, sent forth their huzzahs for Texas!

Colonel Fannin, on account of his wound, was not marched from the fort with the other prisoners. When told that he was to be shot, he heard it unmoved and, giving his watch and money to the officer who was to superintend his execution, he requested that he might not be shot in the head and that his body should be decently buried.

He *was* shot in the head, and his body stripped and pitched into the pile with the others.

The wounded lying in the hospitals were dragged into the fort and shot. Their bodies, like that of Colonel Fannin, were drawn out of the fort about a fourth of a mile and there thrown down.

We now went back to the hospital and resumed our duties. Colonel Garay assured us that we should no longer be confined, but left at large, and that as soon as the wounded got better, we should be released and sent to the United States.

We found that Dr. Field and about a dozen of Fannin's men had been saved. The two men who were concealed under the blanket in the tent were two carpenters by the names of White and Rosenbury, who had done some work for Colonel Garay the day before that pleased him so much that he sent for them in the night and kept them there until the massacre was over.

We continued on attending to the wounded Mexicans for about three weeks. The troops all left Goliad for the east the day after the massacre, leaving only seventy or eighty men to guard the fort and attend to the hospital. Major Miller, by giving his parole that his men would not attempt to escape, obtained for them leave to go at large.

I must not here omit to mention Señora Alvares, whose name ought to be perpetuated to the latest times for her virtues, and whose action, contrasted so strangely with those of her countrymen, deserved to be recorded in the annals of this country and treasured in the heart of every Texan. When she arrived at Copano with her husband, who was one of Urrea's officers, Miller and his men had just been taken prisoner; they were tightly bound with cords so as to completely check the circulation of the blood in their arms, and in this state had been left several hours when she saw them. Her heart was touched at the sight, and she immediately caused the cords to be removed and refreshments given them. She treated them with great kindness, and when, on the morning of the massacre, she learned that the prisoners were to be shot, she so effectually pleaded with Colonel Garay (whose humane feelings revolted at the barbarous order) that, with great personal responsibility to himself and at great hazard at thus going counter to the orders of the then all-powerful Santa Anna, he resolved to save all that he could; and a few of us, in consequence, were left to tell of that bloody day.

Besides those that Colonel Garay saved, she saved by connivance of some of the officers—gone into the fort at night and taken out some, whom she kept concealed until after the massacre. When she saw Dr. Shackelford a few days after and heard that his son was among those sacrificed, she burst into tears and exclaimed, "Why did I not know that you had a son here? I would have saved him at all hazards!" She afterwards showed much attention and kindness to the surviving prisoners, frequently sending messages and presents of provision to them from Victoria.

After her return to Matamoros, she was unwearied in her attention to the unfortunate Americans confined there. She went on to the City of Mexico with her husband. She returned to Matamoros without any funds for her support, but she found many warm friends among those who had heard of and witnessed her extraordinary exertions in relieving the Texas prisoners. It must be remembered that when she came to Texas, she could have considered its people as rebels and heretics, the two classes of all others the most odious to the mind of a pious Mexican; that Goliad, the first town she came to, had been destroyed by them recently, and its Mexican population dispersed to seek refuge where they might; and yet, after everything that had occurred to present the Texans to her view as the worst and most abandoned of men, she became incessantly engaged in contributing to their wants and to save their lives.

Her name deserves to be recorded in letters of gold among those angels who have from time to time been commissioned here by an overruling and beneficent Power to relieve the sorrows and cheer the hearts of men, and who have for that purpose assumed the form of helpless women, that the benefits of the boon might be enhanced by the strong and touching contrast with aggravated evils worked by fiends in human shape, and balm poured on the wounds they make by a feeling of pitying women.

During the ensuing three weeks we could ascertain but little of what was being done by the Mexican army, save the news that came in general terms that Santa Anna was ravaging the whole country and the Texans were flying before him to the Sabine; that Matagorda was taken, and that San Felipe was burned by its own citizens and abandoned on the approach of the army.

April 16th—I now commenced a regular journal. By the request of Colonel Ugartechea, the commandant of Goliad, Dr. Shackelford and myself promised to go up to San Antonio to take in our charge the wounded officers there. The leaving of Goliad, where we had undergone such a variety of fortune, and where every scene recalled such painful remembrances, was truly reviving. We crossed the river and rode through the day over a delightful country covered with patches of shrubbery, now in full verdure; and while our eyes were relieved by reposing on Nature's freshness, the fragrance of the numerous flowers that covered the prairies conveyed exquisite pleasure to another sense, and the balmy breeze seemed to infuse new vigor and give us the feeling of healthful and animated life.

It was while riding along to day that I became struck with the great alteration that six weeks had made in the appearance of my companion, Dr. Shackelford; then an active, hale man of forty, he now seemed at least ten years older and bending under the weight of his sorrows. I remarked the change to him, and he replied that the same idea had occurred to him in regard to myself. The last few weeks, though short, seem an age, and they have not only made us *look* but *feel* older.

Sunday, 17th—Rode about twenty-five miles; prairie high and rolling, with mesquite trees scattered about. We occasionally today had glimpses of the San Antonio timber that winds along like a narrow belt four or five miles to our left. At night, signs of rain; fixed a tent with our saddles, blankets and some bushes. While engaged in camping, it now being dark, a wild goose flying over became dazzled by the campfire and came fluttering down a few feet in front of us. Our soldiers soon despatched it with their swords, and, as we had eaten supper, we dressed it for breakfast.

Monday, 18th—After our goose and a cup of coffee we again resumed our journey. About 10 a.m. we came to a rancho and procured some milk, bread and cheese, which now are the greatest luxuries to us. Passed two or three ranchos today; within a few miles of each we were sure to find two or three hundred head of cattle grazing. At one rancho today we saw goats. Came to a rancho and stopped for the night. These rancheros remind me forcibly of the patriarchs of old in their possessions and simplicity of life. It will not do, however, to trust to people's honesty because of their simplicity, as I found to my cost. I had managed to obtain one change of clothing since being a prisoner, and this night they stole that and left me to travel proudly with all my wealth on my back.

Went on a few miles to a rancho, where we stopped and took breakfast under a large live oak, the branches of which spread out and overshadowed the yard in front of the house. Rode on and crossed the Salado, a beautiful stream of water. Our guide refused to halt here for fear of Indians. Rode on a few miles further and halted to graze our horses on a post oak prairie. After making a fire and cooking, lay down under the shade of a tree; the trunk of another lay near us on the ground, through the crevices of which we discovered the folds of an enormous rattlesnake. We punched him with a stick to drive him out, but His Snakeship only sounded his rattles and drew himself further in. We then set fire to the old trunk and burned up this representative of mankind's first great adversary. After the trunk was burned down, we poked at the cinders and found two large rattlesnakes nicely broiled; the meat looked as nice and smelled as delicious as trout, but we did not feel any inclination to taste it, notwithstanding its flavor.

Rode on, and soon came in sight of the lower Mission, about three miles off to our left—a stately church, left tenantless to waste in the wilderness. Met a company of cavalry en route for Goliad. After passing by two or three other missions about the same distance off, we came in sight of Bexar, and arrived there a little before sunset.

We were conducted to the commandant, General Andrade, who, with several of his officers, we found sitting before one of the houses on one side of the public square.

The moment we entered the town, the whole population—men, women and children—began to flock around us, and by the time we had reached the commandant, it appeared as if the whole town was about us and gazing with the greatest curiosity. Had we been tigers or captive Comanches there could not have been a greater stir. General Andrade, after reading our letters, addressed us, inquiring if we spoke Spanish or French. I answered him in the latter language. He observed that Colonel Ugartechea had written to him as if we had come of our own accord; we were not to be considered as prisoners, but were entitled to our release, and that he had promised us passports to leave as soon as the wounded could get along without assistance.

The general appeared pleased at our coming, and pledged his honor that what had been promised us should be strictly fulfilled. We were then conducted to our quarters. Dr. Shackelford was placed in the house of Don Ramon Musquiz, and I went to that of Angelo Navarro.

April 21st—Yesterday and today we have been round with the surgeons of the place to visit the wounded; and a pretty piece of work "Travis and his faithful few" have made of them. There are about a hundred here now of the wounded. The surgeons inform us that there were four hundred brought into the hospitals the morning they assaulted the Alamo, but I should think from appearances that there were more. Many around the town were crippled there, apparently two or three hundred such; and citizens inform me that three or four hundred have died of their wounds.

We have two colonels, one major and eight captains under our charge who were wounded in the assault. We have taken one ward of the hospital under our charge. Their surgical departments are shockingly conducted; not an amputation performed before we arrived, although there were several cases even now that should have been operated upon from the first; and how many have died from the want of operations is impossible to tell, though it is a fair inference that there have been not a few. There has been scarcely a ball cut out as yet, almost every patient carrying the lead he received on that morning.

In the course of a week after we came to town, a party of Comanches were here. They brought in hams and things to trade to the Mexicans, who made much of them and treated them with a great deal of deference. They are large men and very muscular.

This evening (27th) a family of rancheros coming into town with a cart were attacked two or three miles out by the Tawacana Indians (as they say, but I strongly suspect the Comanches who left two or three days ago). Two or three men and women were killed, one woman dangerously wounded in the stomach, one woman slightly wounded in the back and scalped, and one girl severely wounded. We have taken them in our care and dressed their wounds. I am told that Indians frequently kill people a few miles from town.

We get on very comfortably here. These people show us much respect and courtesy. We meet with much simple and unaffected kindness of heart from the citizens, particularly the females; we are also treated well by the officers. It is evident that they have a high opinion of our skill. If the surgeons that I have seen among them are fair samples of the medical talent in the nation, I can safely say without the least degree of vanity, that they have reason to think well of us. The surgeon of the garrison came for me the other day to visit his wife, who was in the greatest distress and he did not know what to do for her. On going to his house, I found that she had a *tooth ache.* He amputated a leg the day we arrived, and the man died the next. We have as yet amputated but one, and the patient is doing well. There are about a half-dozen more than should have received operations, but now they will die anyhow.

Today I got some clothes from the tailor—the first change that I have been able to get since coming here.

Thursday, 28th—One of the men killed by the Indians yesterday was brought in today, stabbed in several places and scalped. Three or four of our hospital patients died last night. A courier from Goliad but no news of interest. We hear nothing of what is doing on the Brazos or beyond. The officers here appear to know no more than ourselves, and if they do they keep it to themselves. Three Negroes brought in from Tenoxticlan, on the Brazos; a drove of cattle and hogs brought in from the Colorado.

Friday, May 6th—Several Negroes brought in from the Brazos, belonging to Groce and Donahue. Observed this evening, as I walked out, the people collecting in small groups, talking anxiously together and seeming to have some great news that agitated them. I was walking with an officer, and did not think it proper to inquire what it was; but on meeting some Negroes he requested me to ask of them the news. I did so, and they informed me that Santa Anna had lost a battle and was taken.

At first I could not credit it, but reflected a moment on the agitation that was visible among the people, which suggested forcibly to my mind it must be true! And now what torturing anxiety, what racking suspense, till the news is confirmed or falsified! And now what will my countrymen do in the way of reprisal for the outrages perpetrated by this monster? What ought they to do? The few who fell here fell in open fight, it is true, and, fighting to the last, they asked no quarter; and yet does not an order to give no quarter deserve to be retaliated? Does not the killing of Grant and his men, taken by surprise and unable to fight; and the wanton murder of King and his dozen after they could fight no longer; and that worst of outrageous atrocities, the massacre at Goliad in violation of pledged faith and solemn stipulations, deserve, I will not say retaliation, but just vengeance on the author of these enormities, and by whose special order they were perpetrated?

Certainly, now that they have him in their hands, they are bound by every sentiment of regard for their families driven from their homes, and their houses pillaged and burned; by the blood of their brothers and sons, which has flowed like rivers; and by the manes of those whose chivalry led them to meet the foe at the onset, whose bones and ashes lie here and at Goliad, bleaching in the sun, preyed upon by the vulture and the wolf, and when the soil is yet black from their blood that saturated it—they are bound to execute judgment! It may be dangerous for me, but I have faced too many dangers of late for that to influence any sentiments in regard to a principal of right and wrong or a matter of duty and obligation.

Saturday, 7th—The news yesterday seems to be confirmed, and also that there is an armistice, or suspension of hostilities, agreed upon; and the Mexicans are to retreat to this river at least, if not wholly to evacuate the country. Today Don Jose Lombardus, the owner of the wagons and their equipage in this army, showed me a letter he received from Matamoros, dated 21st April, in which was written, "Tomorrow the American prisoners, fourteen in number, are to be shot." Poor fellows! You are to suffer as hundreds of others besides you have done; but in my humble opinion, there will soon be a stop put to this shooting of prisoners, or the tables will be turned with a vengeance. Done Jose is in great trouble about the wagons Santa Anna has lost for him. He fears, and I suspect, truly enough, that the government of Mexico will never pay him for them.

Monday, 9th—Wrote to Dr. Field and Dr. Hale, and also to Dr. Hurtado, at Goliad, by a courier.

Tuesday, 10th—We have many flying reports that tantalize us by their uncertainty and want of credibility: "The Mexican army is coming to San Antonio where it will fortify and send to the interior for reinforcements, so as to take the field early in the fall." "Santa Anna has escaped and regained his liberty"; "Santa Anna and Houston are coming on to San Antonio together in perfect amity." Today we hear, with some appearance of probability, that the Mexican army will not come here, but go to Goliad, and that this place, with all its artillery, ammunition, etc., is to be given up, and that the Texan troops are now on their way here to take possession. Last night there was much riding about town. It is evident that they are agitated about something. Time will show.

Sunday, 15th—Nothing more of news. A Mexican surgeon from Nondova arrived. His name is Nioran, and he seems something more respectable for a surgeon than the others I have seen. Yesterday I strolled over to the Alamo with our hospital captain, Martinez. They are hard at work fortifying. Went along through some of the old gardens; many of the most beautiful flowers are now in bloom; mulberries are ripening, and the fruit of the fig trees begins to appear; but everything of nature's productions looks wild and neglected.

Tuesday, 17th—Dr. Alsbury came into town today with a pass from General Filsaola, now commander-in-chief. The doctor is son-in-law to Angelo Navarro, with whom I live. His wife and sister, together with a Negro, Bowie's, were in the Alamo when it was stormed. He had come in to look after his family and take them off. He gave us all the particulars of the Battle of San Jacinto, the capture of Santa Anna and the retreat of the Mexican army, the number of volunteers pouring into Texas, stimulated thereto by the tale of Fannin and Travis. Now I am truly revived; our cause is prospering and the blood of so many heroes has not been shed in vain.

19th—Dr. Alsbury, in his narrative, related Santa Anna's complimentary speech to General Houston, where he modestly compares himself to Napoleon and Houston to Wellington. There is a sprightly little Frenchman here who is the armorer, and I could not forbear relating to him the anecdote. He sprang up in the greatest excitement. "What!" said he, "Does Santa Anna compare himself to Napoleon, the *foudre*, because he can run about with two or three thousand ragged Indians and take a few mud towns? Does he think that his greatest exploit will bear any comparison to the least thing done by our hero?" He stormed and raved for a considerable time before he could cool down, so indignant was he; and I was much amused at his idea of the comparison.

General Andrade has received orders to destroy the Alamo and proceed to join the main army at Goliad. The troops have hitherto been busy in fortifying the Alamo. They are now as busy as bees—soldiers, convicts and all—tearing down walls, etc. We were promised our passports a few days ago, but there being some difficulty in the way of getting them, finding that the troops were about retreating from here, we have, by means of our friends, Don Jose Lombardus and Don Ramon Musquiz, induced the commandant to leave us here when he goes out, ostensibly in charge of the sick he is obliged to leave behind.

Unembellished Wm. Hunter

[Few Goliad survivor accounts are as exaggerated in narratives by others than that of Judge William L. Hunter. He served as a judge in both Goliad and Refugio Counties and spent the greater part of his life very near the place where his life almost met with a surprise ending in 1836. It is said that he, Dr. Barnard and John C. Duval met and discussed their Goliad experiences. Duval's account of Hunter's escape, in Early Times in Texas, *however, is a far cry from Judge Hunter's telling of it. Duval has Hunter being shot, bayoneted, his throat cut with a butcher knife and his head beaten in with the butt of a gun before getting up to flee! It is remarkable enough without any embellishment but Duval and others seem to have taken certain liberties with the story. The text here is as it was told to J. J. Linn of Victoria for his 1883 book,* Reminiscences of Fifty Years in Texas.*]*

Hunter made an extraordinary escape from the Fannin massacre. He was a member of the New Orleans Greys [San Antonio Greys]. He was shot down at the first fire, and remained for a considerable time unconscious. Upon reviving he could not move his body, as a dead comrade had fallen upon him. Being very weak from the loss of blood, he extricated himself with difficulty, and discovered that he had been stripped of his clothing, retaining only undershirt and drawers. He summoned all his strength for one supreme effort to reach the river, and nearly failed in doing so. He submerged himself in the water, and remained in that position all day.

At night he crossed the river and struck out in an easterly direction. He came to a small stream the next morning, upon the banks of which he remained nearly all day, suffering excruciating pains from his wounds, and being rendered weak from hunger and the loss of blood. He finally made another start, and soon came to another stream, and in following the course of this he came across his own tracks where he had crossed before. He then took down the creek, and came to a house, near the Coletto, where he found some Mexicans who could speak a few words of English, and received of them some clothing and food. These people treated him with the utmost kindness and did all that they could to alleviate his pains. The owner of the jacal, Juan Reynea, had previously lived at the Goliad crossing, but had removed to avoid the unwelcome visits of the soldiers, who were continually passing between Goliad and Victoria.

With the aid of these Good Samaritans, Hunter speedily recovered sufficient strength to resume his journey. Señor Reynea himself accompanied him to the house of Mrs. Margaret Wright, wife of David R. Wright, five miles above Victoria, on the Guadalupe River. This good old mother of Israel died very recently, in the city of Victoria, at the advanced age of eighty-seven years. She nursed Hunter with a mother's care, and sheltered him from the Mexicans until after the Battle of San Jacinto. This statement I have from the lips of Judge Hunter himself, who now resides in Goliad, near the spot or that most terrible episode in his life.

John C. Duval (and Friends)

[John Crittenden Duval (1816-1897) was a Kentucky native who grew into his teen years in the Florida Territory. In his account, he refers, without any implied familial ties, to Capt. Burr H. Duval. The Captain died in the Massacre at Goliad and was John Duval's older brother. By 1845, Duval was serving as a Texas Ranger under Jack Hays, in company with Bigfoot Wallace. The book from which the following is excerpted, Early Times in Texas, *was first published in serial form in* Burke's Weekly *in 1867. It would become his second work in book form when published in 1892 under the same title. After his escape from Goliad, Duval fled for a time with fellow escapees John C. Holliday and Samuel T. Brown. Brown's account of their escape was written 30 years earlier than Duval's and in more haste, and is included in this book. Duval's account generously expounds upon many mentions made in Brown's. Holliday did not, it seems, have the opportunity to write an account. He remained with the army, achieving the rank of Captain in 1837. As part of the Santa Fe Expedition, Holliday was imprisoned in Mexico in 1842. He died aboard the ship taking him to his liberty in Galveston the same year. Holliday Creek and the town of Holliday, Texas are named for Capt. Holliday.]*

...Some time after our arrival at Goliad, information was obtained from some friendly Mexicans that General Santa Anna was preparing to enter Texas at the head of a large army; consequently all idea of invading Mexico was abandoned, and we set to work to render the fortifications around the old missions as defensible as possible. We strengthened the walls in many places, built several new bastions on which artillery was placed in such a way as to command all the roads leading into the town.

Every day we were drilled by our officers for three hours in the morning and two in the afternoon, which, as I have said before was a great bore to me, as I would have preferred passing the time in hunting and fishing. We also deepened the trenches around the walls, and dug a ditch from the fort to the river, and covered it with plank and earth, so that we might obtain a supply of water, if besieged, without being exposed to the fire of the enemy. We were well supplied with artillery and ammunition for the same, and also with small arms, and had beef, sugar and coffee enough to last us for two months—but very little bread.

Some time in February, a Mexican from the Rio Grande arrived at Goliad who informed Col. Fannin that Santa Anna had already or would shortly cross the river into Texas with a large army which would advance in two divisions, one towards Goliad and the other towards the city of San Antonio. Some days afterwards, two or three Texans came in from San Patricio, bringing the news that Capt. Grant and some twenty-five or thirty men stationed at that place, had been surprised by a force of Mexican guerillas and all of them massacred. About this time also a courier from Refugio came in who stated to Col. Fannin that he had been sent by the people of that place, to ask for a detachment of men to escort them to Goliad, as they were daily expecting an attack from the guerillas.

In compliance with this request, Col. Fannin sent Capt. King and his company (about thirty-five men) to act as escort for those families who desired to leave. When Capt. King and his men reached Refugio, they were attacked on the outskirts of the town by a large force of Mexican cavalry, and being hard pressed they retreated into the old mission a strong stone building, at that time encompassed by walls. There they defended themselves successfully, and kept the Mexicans at bay until their artillery came up, when they opened fire upon it with two field pieces which soon breached the walls, and the place was then taken by storm. Capt. King and some seven or eight of his men (the only survivors of the bloody conflict), were captured and led out to a post oak grove north of town, where they were tied to trees and shot. Their bones were found still tied to the trees when the Texan forces re-occupied the place in the summer of '36.

About this time a courier arrived bringing a dispatch from Gen. Houston to Col. Fannin, and it was rumored in camp that the purport of this dispatch was "that Col. Fannin should evacuate Goliad and fall back without delay towards the settlements on the Colorado." But as to the truth of this I cannot speak positively. At any rate Col. Fannin showed no disposition to obey the order if he received it—on the contrary, hearing nothing from Capt. King, although he had sent out three scouts at different times to obtain information of his movements, all of whom were captured and killed, he despatched Maj. Ward with the Georgia Battalion (about one hundred and fifty strong) to his assistance. They were attacked before they reached Refugio by a large force of Mexican cavalry. They made a gallant defense for some time against the vastly superior numbers of the enemy, but at length their ammunition was exhausted and they were compelled to retreat to the timber on the river, where they were surrounded by the Mexican cavalry, and most of them finally captured.

This division of our small force in the face of an enemy so greatly our superior in numbers, was, in my opinion, a fatal error on the part of Col. Fannin.

Hearing nothing either from Capt. King or Major Ward, and satisfied from information obtained by our scouts that a large force of Mexicans was in the vicinity of Goliad, Col. Fannin and his officers held a council of war in which it was determined to evacuate the place and fall back as rapidly as possible towards Victoria on the Guadalupe river. The same day, I believe, or the next after this council of war was held, a courier came in from San Antonio bringing a dispatch, as I was informed, from Col. Travis, to the effect "that he was surrounded in the Alamo by Santa Anna's army, and requesting Col. Fannin to come to his relief without delay. "

Rations for five days and as much ammunition as each man could conveniently carry were immediately issued, and our whole force, including a small artillery company with two or three field pieces, started for San Antonio, crossing the river at the ford a half mile or so above town. After crossing the river and marching a short distance on the San Antonio road, a halt was made and our officers held a consultation, the result of which (I suppose) was the conclusion that we could not reach San Antonio in time to be of any assistance to Col. Travis. At any rate we were marched back to Goliad, recrossing the river at the lower ford.

A few hours after we had got back to our old quarters, a detachment of Mexican cavalry, probably eighty or a hundred strong, showed themselves at a short distance from the fort apparently bantering us to come out and give them a fight. Col. Horton, who had joined us a few days previously with twenty-five mounted men, went out to meet them, but when he charged them they fled precipitately, and we saw them no more that day.

That evening preparations were made to abandon the place; to that end we spiked our heaviest pieces of artillery, buried some in trenches, reserving several field pieces, two or three howitzers and a mortar to take with us on our retreat. We also dismantled the fort as much as possible, burnt the wooden buildings in its immediate vicinity and destroyed all the ammunition and provisions for which we had no means of transportation.

The next morning we bade a final farewell, as we supposed, to Goliad, and marched out on the road to Victoria. We had nine small pieces of ordnance and one mortar, all drawn by oxen as were our baggage wagons. Our whole force comprised about two hundred and fifty men, besides a small company of artillery and twenty-five mounted men under Col. Horton.

We crossed the San Antonio river at the ford below town, and a short distance beyond Manahuilla creek we entered the large prairie extending to the timber on the Coletto, a distance of eight or nine miles. When we had approached within two and a half or three miles of the point where the road we were traveling entered the timber (though it was somewhat nearer to the left) a halt was ordered and the oxen were unyoked from guns and wagons, and turned out to graze. What induced Col. Fannin to halt at this place in the open prairie, I cannot say, for by going two and a half miles further, we would have reached the Coletto creek, where there was an abundance of water and where we would have had the protection of timber in the event of being attacked. I understood at the time that several of Col. Fannin's officers urged him strongly to continue the march until we reached the creek, as it was certain that a large body of Mexican troops were somewhere in the vicinity; but however this may be, Col. Fannin was not to be turned from his purpose, and the halt was made. Possibly he may have thought that two hundred and fifty well-armed Americans under any circumstances would be able to defend themselves against any force the Mexicans had within striking distance, but as the sequel will show the halt at this place was a most fatal one for us. Up to this time we had seen no Mexicans, with the exception of two mounted men, who made their appearance from some timber a long way to our right and who no doubt were spies watching our movements.

At length after a halt of perhaps an hour and a half on the prairie, and just as we were about to resume our march for the Coletto, a long dark line was seen to detach itself from the timber behind us, and another at the same time from the timber, to our left. Someone near me exclaimed, "Here come the Mexicans!" and in fact, in a little while, we perceived that these dark lines were men on horseback, moving rapidly towards us. As they continued to approach, they lengthened out their columns, evidently for the purpose of surrounding us, and in doing so displayed their numbers to the greatest advantage. I thought there were at least ten thousand (having never before seen a large cavalry force), but in reality there were about a thousand besides several hundred infantry (mostly Carise Indians).

In the meantime we were formed into a "hollow square" with lines three deep, in order to repel the charge of the cavalry, which we expected would soon be made upon us. Our artillery was placed at the four angles of the square, and our wagons and oxen inside. Our vanguard under Col. Horton, had gone a mile or so ahead of us, and the first intimation they had of the approach of the enemy was hearing the fire of our artillery when the fight began. They galloped back as rapidly as possible to regain our lines, but the Mexicans had occupied the road before they came up and they were compelled to retreat. The Mexicans pursued them beyond the Coletto, but as they were well mounted they made their escape.

The loss of these mounted men was a most unfortunate one for us. Had they been with us that night after we had driven off the Mexicans, we would have had means of transportation for our wounded, and could easily have made our retreat to the Coletto.

When the Mexicans had approached to within half a mile of our lines they formed into three columns, one remaining stationary, the other two moving to our right and left, but still keeping at about the same distance from us. Whilst they were carrying out this maneuver, our artillery opened upon them with some effect, for now and then we could see a round shot plough through their dense ranks. When the two moving columns, the one on the right and the one on the left were opposite to each other, they suddenly changed front and the three columns with trumpets braying and pennons flying, charged upon us simultaneously from three directions.

When within three or four hundred yards of our lines our artillery opened upon them with grape and canister shot, with deadly effect, but still their advance was unchecked, until their foremost ranks were in actual contact in some places with the bayonets of our men. But the fire at close quarters from our muskets and rifles was so rapid and destructive, that before long they fell back in confusion, leaving the ground covered in places with horses and dead men.

Capt. Duval's company of Kentucky riflemen and one or two small detachments from other companies formed one side of our "square," and in addition to our rifles, each man in the front rank was furnished with a musket and bayonet to repel the charge of cavalry. Besides my rifle and musket I had slung across my shoulders an "escopeta," a short light blunderbuss used by the

Mexican cavalry, which I had carried all day in expectation of a fight, and which was heavily charged with forty "blue whistlers" and powder in proportion.

It was my intention only to fire it when in a very tight place, for I was well aware it was nearly as dangerous behind it as before. In the charge made by the Mexican cavalry they nearly succeeded in breaking our lines at several places, and certainly they would have done so had we not taken the precaution of arming all in the front rank with the bayonet and musket. At one time it was almost a hand-to-hand fight between the cavalry and our front rank, but the two files in the rear poured such a continuous fire upon the advancing columns, that, as I have said, they were finally driven back in disorder. It was during this charge and when the Mexican cavalry on our side of the square were in a few feet of us, that I concluded that I had got into that tight place and that it was time to let off the "scopet" I carried.

I did so, and immediately I went heels over head through both ranks behind me. One or two came to my assistance supposing no doubt I was shot (and in truth I thought for a moment myself that a two ounce bullet had struck me) but I soon rose to my feet and took my place in the line again just as the cavalry began to fall back. Now, I don't assert that it was the forty blue whistlers I had sent among them from my scopet that caused them to retreat in confusion. I merely mention the fact that they did fall back very soon I after I had let off the blunderbuss among them. My shoulder was black and blue from the recoil for a month afterwards. When I took my place in the line again, I never looked for my scopet, but contented myself while the fight lasted with my rifle.

The Mexicans had no doubt supposed they would be able break our lines at the first charge, and were evidently much disconcerted by their failure to do so; for although they reformed their broken columns and made two more attempts to charge us, they were driven back as soon as they came within close range of our small arms.

When they were satisfied that it was impossible for them to break our lines, the cavalry dismounted and, surrounding us in open order, they commenced a fusillade upon us with their muskets and escopetas, but being very poor marksmen, most of their bullets passed harmlessly over our heads. Besides, this was a game at which we could play also, and for every man killed or wounded on our side I am confident that two or three Mexicans fell before the deadly fire from our rifles. But there were with the Mexicans probably a hundred or so Carise Indians, who were much more daring, and withal better marksmen. They boldly advanced to the front and, taking advantage of every little inequality of the ground and every bunch of grass that could afford them particular cover, they would crawl up closely and fire upon us, and now and then the discharge of their long single-barrel shotguns was followed by the fall of someone in our ranks.

Captain Duval, who was using a heavy Kentucky rifle and was known to be one of the best marksmen in his company, was requested to silence these Indians. He took a position near a gun carriage, and whenever one of the Indians showed his head above the tall grass, it was perforated with an ounce rifle ball, and after four shots they were seen no more. At the moment that he fired the last shot, Captain Duval had one of the fingers of his right hand taken off by a musket ball. When the Mexicans quit the field, we examined the locality where these Indians had secreted themselves and found the four lying closely together, each one with a bullet hold through his head.

At the commencement of the fight a little incident of a somewhat ludicrous character occurred. We had some five or six Mexican prisoners (the couriers of the old padre, captured at Carlos' Ranch). These we had placed within the square, when the fight began, for safekeeping, and in an incredibly short time, with picks and shovels, they dug a trench deep enough to "hole" themselves, where they lay perdu and completely protected from bullets. I for one, however, didn't blame them, as they were non-combatants, and besides to tell the truth when bullets were singing like mad hornets around me, and men were struck down near me, I had a great inclination to "hole up" myself and draw it in after me.

The fight continued in a desultory kind of way, until near sunset, when we made a sortie upon the dismounted cavalry, and they hastily remounted and fell back to the timber to our left, where, as soon as it was dark, a long line of fires indicated the position of their encampment.

That night was anything but rest for us, for anticipating a renewal of the fight the next morning, all hands were set to work digging entrenchments and throwing up embankments, and at this we labored unceasingly till nearly daylight. We dug four trenches enclosing a square large enough to contain our whole force, throwing the earth on the outside, on which we placed our baggage and everything else available, that might help to protect us from the bullets of the enemy.

Before we began this work, however, Col. Fannin made a short speech to the men, in which he told us "that in his opinion, the only way of extricating ourselves from the difficulty we were in, was to retreat after dark to the timber on the Coletto, and cut our way through the enemy's lines should they attempt to oppose the movement." He told us there was no doubt we would be able to do this, as the enemy had evidently been greatly demoralized by the complete failure of the attack they had made upon us. He said, moreover, that the necessity for a speedy retreat was the more urgent, as it was more than probable that the Mexicans would be heavily reinforced during the night. He concluded by saying that if a majority were in favor of retreating, preparations would be made to leave as soon as it was dark enough to conceal our movements from the enemy. But we had about seventy men wounded (most of them badly) and, as almost everyone had some friend or relative among them, after a short consultation upon the subject, it was unanimously determined not to abandon our wounded men, but to remain with them and share their fate, whatever it might be.

Our loss in the Coletto fight was ten killed and about seventy wounded (Col. Fannin among the latter), and most of them badly, owing to the size of the balls thrown by the Mexican escopetas, and the shotguns of the Indians. The number of our casualties was extremely small considering the force of the enemy and the duration of the fight, which began about three o'clock and lasted till nearly sunset. I can only account for it by the fact that the Mexicans were very poor marksmen, and that their powder was of a very inferior quality. There was scarcely a man in the whole command who had not been struck by one or more spent balls, which, in place of mere bruises would have inflicted dangerous or fatal wounds if the powder used by the Mexicans had been better.

I can never forget how slowly the hours of that dismal night passed by. The distressing cries of our wounded men begging for water when there was not a drop to give them, were continually ringing in my ears. Even those who were not wounded, but were compelled to work all night in the trenches, suffered exceedingly with thirst. Even after we had fortified our position as well as we could, we had but little hopes of being able to defend ourselves, should the Mexicans as we apprehended, receive reinforcements during the night, for we had but one or two rounds of ammunition left for the cannon, and what remained for the small arms was not sufficient for a protracted struggle.

Some time during the night it was ascertained that three of our men (whose names I have forgotten) had deserted, and shortly afterwards, as a volley of musketry was heard between us and the timber on the Coletto, they were no doubt discovered and shot by the Mexican patrol.

Daylight at last appeared, and before the sun had risen we saw that the Mexican forces were all in motion, and evidently preparing to make another attack upon us. When fairly out of the timber, we soon discovered that they had been heavily reinforced during the night. In fact, as we subsequently learned from the Mexicans themselves, a detachment of seven hundred and fifty cavalry and an artillery company had joined them shortly after their retreat to the timber. In the fight of the previous day they had no cannon.

They moved down upon us in four divisions, and when within five or six hundred yards, they unlimbered their field pieces (two brass nine pounders) and opened fire upon us. We did not return their fire, because as I have said, we had only one or two rounds of ammunition left for our cannon, and the distance was too great for small arms. Their shot, however, all went over us, and besides, the breastworks we had thrown up would have protected us, even if their guns had been better aimed. We expected momentarily that the cavalry would charge us, but after firing several rounds from their nine pounders, an officer accompanied by a soldier bearing a white flag, rode out towards us, and by signs gave us to understand that he desired a "parley." Major Wallace and several other officers went out and met him about halfway between our "fort" and the Mexican lines.

The substance of the Mexican officer's communication (as I understood at the time) was to the effect "that Gen. Urrea, the commander of the Mexican forces, being anxious to avoid the useless shedding of blood (seeing we were now completely in his power), would guarantee to Col. Fannin and his men, on his word of honor as an officer and gentleman, that we would be leniently dealt with provided we surrendered at *discretion*, without further attempt at hopeless resistance." When this message was delivered to Col. Fannin, he sent word back to the officer "to say to Gen. Urrea, it was a waste of time to discuss the subject of surrendering at *discretion*—that he would fight as long as there was a man left to fire a gun before he would surrender on such terms."

A little while afterwards the Mexicans again made a show of attacking us, but just as we were expecting them to charge, Gen. Urrea himself rode out in front of his lines accompanied by several of his officers and the soldier with the white flag. Col. Fannin and Major Wallace went out to meet them, and the terms of capitulation were finally agreed upon, the most important of which was, that we should be held as prisoners of war until exchanged, or liberated on our parole of honor not to engage in the war again—at the option of the Mexican commander in chief. There were minor articles included in it, such as that our side arms should be retained, etc.

When the terms of capitulation had been fully decided upon, Gen. Urrea and his secretary and interpreter came into our lines with Col. Fannin, where it was reduced to writing, and an English translation given to Col. Fannin which was read to our men. I am thus particular in stating what I know to be the facts in regard to this capitulation, because I have seen it stated that Gen. Santa Anna always asserted there was no capitulation, and that Col. Fannin surrendered at discretion to Gen. Urrea. This assertion I have no doubt was made to justify as far as possible his order for the cold-blooded murder of disarmed prisoners. Gen. Urrea, I believe, never denied the fact of the capitulation, and I have been informed, when the order was sent him by Santa Anna to execute the prisoners, he refused to carry it into effect, and turned over the command to a subaltern.

I have always believed myself that Gen. Urrea entered into the capitulation with Col. Fannin in good faith, and that the massacre of the prisoners, which took place some days afterwards, was by the express order of Santa Anna, and against the remonstrances of Gen. Urrea. If Gen. Urrea had intended to act treacherously, the massacre, in my opinion, would have taken place as soon as we had delivered up our arms, when we were upon an open prairie, surrounded by a large force of cavalry, where it would have been utterly impossible for a single soul to have escaped, and consequently he could then have given to the world his own version of the affair without fear of contradiction.

I have said nothing as yet of the Mexican loss in the fight and I cannot do so with any certainty of my own knowledge; but there is no doubt it was much greater than ours. They told us after we had surrendered that we had killed and wounded several hundred. Dr. Joseph Barnard, our assistant surgeon, who was saved from the massacre to attend their wounded, told me afterwards that he was confident we had killed and wounded between three and four hundred, and his opportunities for forming a correct estimate of the number were certainly better than those of anyone else.

After our surrender we were marched back to Goliad, escorted by a large detachment of cavalry, and there confined within the walls surrounding the old mission.

Among the Mexican officers there was a lieutenant by the name of Martinez, who had been educated at a Catholic college in Kentucky, where he had been a roommate of a member of Capt. Duval's company, by the name of Brashear. Every day whilst we were prisoners he used to come and talk with Brashear, and professed his great regret to find him in such a situation, but he never gave him the slightest intimation of the treacherous designs of the Mexicans, nor, as far as I know, made the least effort to save his college roommate.

A day or so after our return as prisoners to Goliad, Maj. Ward and his battalion, or rather those who survived the engagement with the Mexicans near Refugio, were brought in and confined with us, within the walls enclosing the old mission; and also a company of about eighty men under the command of Maj. Miller, who had been surprised and captured at Copano just after they had landed from their vessel. These men were also confined with us, but kept separate from the rest, and to distinguish them, each had a white cloth tied around one of his arms. At the time, I had no idea why this was done, but subsequently I learned the reason.

The morning of the sixth day after our return to Goliad, whether the Mexicans suspected we intended to rise upon the guard, or whether they merely wished to render our situation as uncomfortable as possible, I know not, but at any rate from that time we were confined in the old mission, where we were so crowded we had hardly room to lie down at night. Our rations too, about that time, had been reduced to five ounces of fresh beef a day, which we had to cook in the best way we could and eat without salt.

Although, thus closely confined and half-starved, no personal indignity was ever offered to us to my knowledge, except on two occasions. Once a Mexican soldier pricked one of our men with his bayonet, because he did not walk quite fast enough to suit him, whereupon he turned and knocked the Mexican down with his fist. I fully expected to see him roughly handled for this "overt act," but the officer in command of the guard, who saw the affair, came up to him and, patting him on the shoulder, told him he was "muy bravo," and that he had served the soldier exactly right. At another time one of our men was complaining about his ration to the officer of the guard, who ordered one of the soldiers to collect a quantity of bones and other offal lying around, and throwing them on the ground before the man, said, "There, eat as much as you want—good enough for Gringoes and heretics."

One day an officer who was passing asked me some question in Spanish, and when I answered him in Spanish, he took a seat by me and talked with me for some time. He asked me a great many questions about the United States, our form of government, the number of our regular army, what state I came from and what induced me to come to Texas, etc., to all of which I frankly answered. He expressed much astonishment at the correctness of my pronunciation and asked where I had learned to speak Spanish, saying he was sure I had not learnt the language among the Mexicans. I told him I had studied Spanish under a teacher of modern languages at a Catholic institution in Kentucky. He then asked if I was a Catholic myself, and when I told him I was not, he seemed disappointed, and tried in various ways to get some sort of admission from me that I had more faith in the Catholic religion than any other.

The talk I had with this officer made but little impression upon me at the time, but I have since thought that, on account of my youth or because I had in some way gained his favor, he was desirous of an excuse or pretext to save me from the fate he probably knew was in store for us. I know that several of our men were saved from the massacre, for no other reason that I am aware of, than that they *professed* to be members of the Catholic church. Several times afterwards the officer above mentioned came to talk with me, and he insisted I was a Catholic if I would but own it; but I strenuously denied "the soft impeachment" to the last. If I had suspected his object in getting me to admit that I was a Catholic, it is probable I might have sought temporal as well as eternal safety the bosom of the church. It would have been very easy for me to have passed for a "good Catholic," for Catholicism (at least among the lower class of Mexicans) consists mainly in knowing how to make the sign of the cross, together with unbounded reverence first, for the Virgin Mary, and secondly, for the saints generally—and the priests. But I did not suspect the object this officer had in view when he tried to make a convert of me to the true faith, and I am afraid I have lost the only chance I shall ever have of becoming a "good Catholic. "

On the morning of the 27th of March, a Mexican officer came to us and ordered us to get ready for a march. He told us we were to be liberated on parole and that arrangements had been made to send us to New Orleans on board of vessels then at Copano. This, you may be sure, was joyful news to us, and we lost no time in making preparations to leave our uncomfortable quarters. When all was ready we were formed into three divisions and marched out under a strong guard. As we passed by some Mexican women who were standing near the main entrance to the fort, I heard them say "pobrecitos" (poor fellows), but the incident at the time made but little impression on my mind.

One of our divisions was taken down the road leading to the lower ford of the river, one upon the road to San Patricio, and the division to which my company was attached, along the road leading to San Antonio. A strong guard accompanied us, marching in double files on both sides of our column. It occurred to me that this division of our men into three squads, and marching us off in three directions, was rather a singular maneuver, but still I had no suspicion of the foul play intended us.

When about half a mile above town, a halt was made and the guard on the side next the river filed around to the opposite side. Hardly had this maneuver been executed, when I heard a heavy firing of musketry in the directions taken by the other two divisions. Someone near me exclaimed "Boys! They are going to shoot us!" and at the same instant I heard the clicking of musket locks all along the Mexican line. I turned to look, and as I did so, the Mexicans fired upon us, killing probably one hundred out of the one hundred and fifty men in the division.

We were in double file and I was in the rear rank. The man in front of me was shot dead, and in falling he knocked me down. I did not get up for a moment, and when I rose to my feet, I found that the whole Mexican line had charged over me, and were in hot pursuit of those who had not been shot and who were fleeing towards the river about five hundred yards distant. I followed on after them, for I knew that escape in any other direction (all open prairie) would be impossible, and I had nearly reached the river before it became necessary to make my way through the Mexican line ahead. As I did so, one of the soldiers charged upon me with his bayonet (his gun I suppose being empty). As he drew his musket back to make a lunge at me, one of our men coming from another direction, ran between us, and the bayonet was driven through his body. The blow was given with such force that, in falling, the man probably wrenched or twisted the bayonet in such a way as to prevent the Mexican from withdrawing it immediately. I saw him put his foot upon the man, and make an ineffectual attempt to extricate the bayonet from his body. One look satisfied me, as I was somewhat in a hurry just then, and I hastened to the bank of the river and plunged in.

The river at that point was deep and swift, but not wide, and being a good swimmer, I soon gained the opposite bank, untouched by any of the bullets that were pattering in the water around my head. But here I met with an unexpected difficulty.

The bank on that side was so steep I found it was impossible to climb it, and I continued to swim down the river until I came to where a grapevine hung from the bough of a leaning tree nearly to the surface of the water. This I caught hold of and was climbing up it hand-over-hand, sailor fashion, when a Mexican on the opposite bank fired at me with his escopeta, and with so true an aim, that he cut the vine in two just above my head, and down I came into the water again. I then swam on about a hundred yards further, when I came to a place where the bank was not quite so steep, and with some difficulty I managed to clamber up.

The river on the north side was bordered by timber several hundred yards in width, through which I quickly passed and I was just about to leave it and strike out into the open prairie, when I discovered a party of lancers nearly in front of me, sitting on their horses, and evidently stationed there to intercept anyone who should attempt to escape in that direction. I halted at once under cover of the timber, through which I could see the lancers in the open prairie, but which hid me entirely from their view.

Whilst I was thus waiting and undecided as to the best course to pursue under the circumstances, I saw a young man by the name of Holliday, one of my own messmates, passing through the timber above me in a course that would have taken him out at the point directly opposite to which the lancers were stationed. I called to him as loudly as I dared and fortunately, being on the "qui vive," he heard me, and stopped far enough within the timber to prevent the lancers from discovering him. I then pulled off a fur cap I had on, and beckoned to him with it. This finally drew his attention to me, and as soon as he saw me he came to where I was standing, from whence, without being visible to them, the lancers could be plainly seen.

A few moments afterwards we were joined by a young man by the name of Brown, from Georgia, who had just swam the river, and had accidentally stumbled on the place where Holliday and I were holding a council of war as to what was the best course to pursue. Holliday, although a brave man, was very much excited, and had lost to some extent his presence of mind, for he proposed we should leave the timber at once and take the chances of evading the lancers we saw on the prairie. I reasoned with him on the folly of such a proceeding, and told him it would be impossible for us to escape in the open prairie from a dozen men on horseback. "But," said Holliday, "the Mexicans are crossing the river behind us, and they will soon be here."

"That may be," I replied, "but they are not here yet, and in the meantime something may turn up to favor our escape." Brown took the same view of the case I did, and Holliday's wild proposition to banter a dozen mounted men for a race on the open prairie was laid upon the table.

Whilst we were debating this (to us) momentous question, some four or five of our men passed out of the timber before we saw them, into the open prairie, and when they discovered the lancers it was too late. The lancers charged upon them at once, speared them to death and then, dismounting, robbed them of such things as they had upon their persons. From where we stood the whole proceeding was plainly visible to us, and as may be imagined, it was not calculated to encourage any hopes we might have had of making our escape. However, after the lancers had plundered the men they had just murdered, they remounted, and in a few moments set off in a rapid gallop down the river to where it is probable they had discovered other fugitives coming out of the timber. We at once seized the opportunity thus afforded us to leave the strip of timber which we knew could give us shelter but for a few moments longer, and started out, taking advantage of a shallow ravine which partially hid us from view. We had scarcely gone two hundred yards from the timber, when we saw the lancers gallop back and take up their position at the same place they had previously occupied. Strange to say, however, they never observed us, although we were in plain view of them for more than a quarter of a mile, without a single brush or tree to screen us.

We traveled about five or six miles and stopped in a thick grove to rest ourselves, where we stayed until night. All day long we heard at intervals irregular discharges of musketry in the distance, indicating, as we supposed, where fugitives from the massacre were overtaken and shot by the pursuing parties of Mexicans.

As the undergrowth was pretty dense in the grove where we had stopped, we concluded the chances of being picked up by one of these pursuing parties would be greater if we traveled on than if we remained where we were, and we determined to lie by until night. In talking the matter over and reflecting upon the many narrow risks we had run in making our escape, we came to the conclusion that in all probability we were the only survivors of the hundreds who had that morning been led out to slaughter; although in fact as we subsequently learned, twenty-five or thirty of our men eventually reached the settlements on the Brazos.

Drs. Shackelford and Barnard, our surgeons, were saved from the massacre to attend upon Mexicans wounded in the fight on the Coletto, and when their forces retreated from Goliad after the Battle of San Jacinto, these were taken to San Antonio, where they were ultimately liberated. Our own wounded men, or rather those of them that survived up to the time of the massacre, were carried out into the open square of the fort, and there cruelly butchered by the guard. Capt. Miller and his men were saved, because, as I was subsequently informed, they had been captured soon after they landed from their vessel, without any arms, and of course without making any resistance.

Col. Fannin, who was confined to his quarters by a wound he had received at the fight on the Coletto, soon after the massacre of his men, was notified to prepare for immediate execution. He merely observed that he was ready then, as he had no desire to live after the cold-blooded, cowardly murder of his men. He was thereupon taken out to the square by a guard, where he was seated on a bench and his eyes blindfolded. A moment before the order to fire was given, I was told (though I cannot vouch for the truth of the statement) he drew a fine gold watch from his pocket and, handing it to the officer in command of the guard, requested him as a last favor to order his men to shoot him in the breast and not in the head. The officer took the watch, and immediately ordered the guard to fire at his head. Col. Fannin fell dead and his body was thrown into one of the ravines near the fort. Thus died as brave a son of Georgia as ever came from that noble old state…

As soon as it was dark we left our hiding place and set out in a northeasterly direction, as nearly as we could determine, and traveled until daylight, when we stopped an hour or so in a grove to rest. We then proceeded on our course again till near sunset, when we encamped in a thick mot of timber without water. An unusually cold norther for the season of the year was blowing, and a steady drizzling rain was falling when we stopped. Brown, who had pulled off his coat and shoes before he swam the San Antonio River, suffered severely, and I was apprehensive, should we be exposed all night to such weather without a fire, that he would freeze to death. I had a little tinder box in my pocket containing a flint and steel, but all the tinder there was in it was a small piece not much larger than a pin head.

This I carefully placed on a batch of cotton taken from the lining of my fur cap, and after many unsuccessful efforts I managed at last to ignite it. With this we started a fire, and then the first thing I did was to tear off a portion from my drawers, which I partially burned, thus securing a good supply of tinder for future use. Before going to sleep we collected fuel enough to last until daylight, with which we occasionally replenished the fire so that we passed the night in tolerable comfort.

The next morning, Brown, who as I have previously stated, had pulled off his coat and shoes and thrown them away when he swam the river, found himself so sore and crippled he was unable to travel. The prairie we had passed over the day before, had been recently burned off and the sharp points of the stubble had lacerated his naked feet dreadfully. It was evident he could not go on without some sort of covering for his feet. I cut off the legs of my boots, and with a pair of scissors which he happened to have in his pocket, and some twine, I contrived to make him a pair of sandals, such as I had seen worn by Mexican soldiers. After thus shoeing him (by way of remuneration, I suppose,) Brown separated the two blades of the scissors and gave me one of them, which was of great service to me, for by whetting it on stones I gave it an edge, and it answered pretty well in place of a knife.

The grove of timber in which we had passed the night, covered perhaps an acre of ground, and just outside of it, there was a strip of sandy soil almost bare of grass. In the morning when we left the grove we observed a good many fresh moccasin tracks which must have been made during the night by a party of Indians, who probably had been drawn to the locality by the light from our fire. Why they did not attack us I cannot imagine, unless it was because they were ignorant of our number and that we were without arms. At any rate, but for their tracks in the sand we would not have known they had been around our camp during the night.

The next morning we set out, as we supposed, in the direction we had traveled the day before, and in about one hour we came to some timber, bordering upon what I thought was one of the branches of the Coletto creek. Here we laid ourselves down on the grass to rest for a few moments, and scarcely had we done so when a party of ten Mexican lancers made their appearance, riding along a trail that ran within fifty yards of where we were lying. As luck would have it, just as they came opposite to where we were, they met another soldier and stopped to have a talk with him. For nearly an hour, it seemed to me, but in fact, I suppose, for only a few minutes, they sat on their horses conversing together within a few paces of where we were lying, and without a single bush or tree intervening to hide us from their view, but fortunately they never looked toward us or we would inevitably have been discovered. At length they rode on, and we were very glad when we lost sight of them behind a point of timber.

The weather still continued cloudy and drizzly, and not being able to see the sun we had nothing to guide us, and in consequence were doubtful as to whether or not we were pursuing the right course. However, we traveled on until night, and again encamped in a thick grove of timber. Having eaten nothing since we left Goliad, and only a small piece of beef for two days previously, we had begun to suffer severely from the pangs of hunger. Game we had everywhere seen in the greatest abundance, but having no guns, the sight of herds of deer and flocks of wild turkeys, suggestive as they were to our minds of juicy steaks and roasts, only served to aggravate the cravings of our appetite. It was at a season of the year, too, when no berries or wild fruits were to be found, and the pecans and other nuts had fallen and been destroyed by wild hogs, deer and other animals. But in spite of our hunger we slept pretty well on our beds of dry leaves, except that we were occasionally aroused from our slumbers by the howling of wolves, which were sometimes so impudent as to approach within a few paces of the fire about which we were lying.

In the morning the weather was still cloudy and cold, and we set out again upon our travels. Holliday being by several years the oldest of our party, had heretofore taken the lead to which Brown and I had made no opposition, but after an hour or so I was convinced he was leading us in the wrong direction, and in this opinion I was confirmed when in a little while we came to a creek I was pretty sure was the Manahuilla, the same we had crossed the day after leaving Goliad. I told Holliday I was confident he was taking the back track, but he thought not, and so we kept on until toward evening, when we came to several groves of live oak timber which I remembered having seen when hunting in the vicinity of Goliad. Holliday, however, had but little faith in my recollections of the locality, and proposed that Brown and myself should wait in one of these groves until he reconnoitered the country ahead, and we consented to do so.

In about an hour he returned and told us that he had been in sight of Goliad, and that he had distinctly heard the beating of drums and the bugle calls of cavalry in the town. We felt very much discouraged, as may well be supposed, to find ourselves, after traveling so long, almost at the same point we had started from; but it was useless to repine, and we set out again in the right direction, Holliday, as usual, leading the way. After an hour or so I found that Holliday was gradually turning his course toward Goliad again. Time with us was too precious to be wasted. I came to a halt and told Holliday I would follow him no farther. He insisted he was going the right direction, and I as positively that he was going directly contrary to the course we ought to pursue. He was obstinate, and so was I. Holliday, I knew, had been born and raised in a city whilst I had lived the greater part of my life on the frontier, and had been accustomed to the woods ever since I was old enough to carry a gun. Besides, I knew that I possessed to a considerable degree what frontiersmen call "hog-knowledge," by which is meant a kind of instinctive knowledge that enables some people to steer their way through pathless woods and prairies without a compass or any landmarks to guide them.

I therefore told Holliday that if he persisted in traveling in the direction he was going, we would certainly have to part company, although I was very loath to do so under the circumstances. Thereupon and without further parley, I turned and took the opposite course to the one we had been traveling. Brown, who made no pretensions to being a woodman, followed me, for the reason, I suppose, that he had lost confidence in Holliday as a guide and thought possibly I might do better. Holliday remained standing where we had left him, apparently undetermined what to do, until we had gone perhaps a hundred yards, when he turned and followed us. As he came up, he merely said that he would rather go wrong than part company, and no allusion afterwards was made to the subject—but from that time on, I always took the lead as a matter of *course.*

Recrossing the Manahuilla Creek, and night coming on shortly afterwards, we encamped by the side of a pool of water in a thick island of timber. By this time, we were suffering greatly with hunger. Nevertheless I slept soundly through the night, although in my slumbers I was constantly tantalized by dreams of juicy steaks, hot biscuits and butter, etc., which always mysteriously disappeared when I attempted to grab them.

The next morning we again took our course across the prairie, but owing to the rank growth of grass with which in many parts it was covered, and our increasing weakness, our progress was slow and painful. On the way, Holliday found about a dozen wild onions, which he divided with Brown and myself, but the quantity for each was so small that it seemed only to aggravate the pangs of hunger. During the day, we saw in the distance several parties of Mexicans or Indians, we could not tell which, as they only came near enough for us to see that they were men on horseback.

That night we again encamped in a strip of woods bordering a small creek, but we slept very little on account of our sufferings from hunger, which had now become excruciating. The next morning Brown was so weak he could scarcely walk two hundred yards without stopping to rest, nevertheless we went on as fast as we could travel. A part of the way was over high rolling prairie, on which no water could be found, and the pangs of thirst were added to those of hunger, until alleviated by the juice of some "Turks heads" which we found growing on the top of a pebbly knoll. These plants are, I believe, a species of the cactus, about the size of a large turnip, grow on top of the ground, and are protected on the outside by a number of tough, horny prickles. The inside is filled with a spongy substance, which when pressed yields a quantity of tasteless juice that answers as a tolerable substitute for water.

The evening of the fifth day after leaving Goliad, we descried a long line of timber ahead of us, and just before sunset we came to a large stream, which from my knowledge of the geography of the country I was sure must be the Guadalupe. At the point where we struck it, the prairie extended up to the bank, which was high and very steep. A few hundred yards above us we saw a cow and her calf grazing near the edge of the bluff, and approaching them cautiously we attempted to drive them over it, hoping that one or the other would be disabled or killed by the fall, but after several ineffectual efforts to force them to take the leap, they finally broke through our line and made their way to the prairie, taking with them some steaks we stood very much in need of.

Completely exhausted by our exertions, and suffering extremely from hunger, we looked around for a suitable place to camp as it was now nearly night, and coming to a pit or sink twelve or fourteen feet deep, which would protect us from the cold wind blowing at the time, we built a fire at the bottom, laid down upon the leaves, and in a little while we all went to sleep. How long I had slept I do not know, but I was at length aroused from my slumbers by a rattling among the sticks and dry leaves above me, and looking up I discovered a wild sow with her litter of pigs coming down the almost perpendicular bank of the sink. I silently grasped a billet of wood lying near me, and awaited their approach. The old sow came on, totally unsuspicious that three ravenous chaps were occupying her bed at the bottom (for by this time our fire had burnt out), and when she and her pigs were in striking distance I suddenly sprang up and began a vigorous assault upon the pigs.

The noise aroused Brown and Holliday, and comprehending at once the state of affairs they sprang to my assistance, and before the sow and her pigs could make their escape up the steep sides of the pit we had "bagged" five of the latter. We made a desperate attack on the old sow also, but weak as we were from starvation, and with our inefficient weapons, she routed us completely, leaving us however in possession of the field and the "spoils of war."

We immediately started our fire again, and with no other preparation than a slight roasting on the coals, enough to singe off their hair, we very expeditiously disposed of the five pigs we had killed—nearly a pig and a half for each one, but then you must remember that they were small sucking pigs, and that we had not had a mouthful to eat for five days except a handful of wild onions. Greatly refreshed by our supper of scorched pig, we laid down again upon the leaves at the bottom of the sink, and slept soundly until the sun was an hour or so high.

As soon as we awoke, we left the sink and went out to make a reconnaissance of the river, to see what the chances were for crossing it. Though not very wide at that point, we soon perceived we had a difficult job to undertake, for the river was much swollen by recent rains, and its turbid waters were rushing along at a rapid rate. Holliday and I were both good swimmers, and I felt sure we could reach the opposite bank safely, but I had my doubts about Brown. He was a poor swimmer, and consequently was timid in water. However, there was no alternative but to make the attempt, and we therefore stripped off our clothes, tied them in a bundle on our heads to keep them as dry as possible, and plunged in the turbid flood.

Holliday and I soon reached the opposite bank, but hardly had we done so when I heard Brown cry out for help, and looking back I saw that he was still some distance from the shore, and evidently just on the eve of going under. At the very point where I landed there happened to be a slab of dry timber lying near the water, which I instantly seized, and swimming with it to the place where Brown was struggling to keep his head above the surface, I pushed the end of the slab to him, which he grasped and to which he held on with the usual tenacity of a drowning man, and with the assistance of Holliday I at last got him to the shore and dragged him out of the water. It was very fortunate for Brown that Holliday and I, between us, had taken his clothes, as otherwise no doubt he would have lost them all.

Continuing our course, we passed though a heavily timbered bottom more than a mile wide, and then came to a large prairie in which we saw many herds of deer and some antelopes. The antelope is a beautiful animal about the size of a deer, but much more fleet. They do not run as deer do, by springs or bounds, but evenly, like the horse. Their horns consist of two curved shafts, with a single prong to each. A man on a good saddle horse can easily overtake a fat deer on the prairie, but it would require a thoroughbred racer with a light rider to come up with an antelope.

We also saw today a party of Indians on horses, but we eluded them by concealing ourselves in some tall grass that grew in the bottom of a ravine. About dusk we came to the timber on the farther side of the prairie, in which we encamped under the spreading branches of a live oak tree.

Next morning we continued our route, and after passing through some open post oak woods, we came to a small stream not more than knee deep, and of course easily forded. Crossing this stream, we went through more post oak woods, and then entered another large prairie, and it was late in the evening, owing to the difficulty of making our way through tall and tangled grass, before we reached the timber on the opposite side, where we encamped in a little open space surrounded by a dense growth of underwood. Here we made a fire, and slept soundly till morning.

As soon as daylight appeared we were off again, and passing through a skirt of woods we came to another small stream, which was also fordable. Crossing it, we entered a large prairie, on the opposite side of which a long line of timber was dimly visible in the distance. All day long, stopping occasionally to rest, we toiled through the matted grass with which this prairie was covered, and just at sunset we came to the woods we had seen, where we encamped near a pool of water. Whilst collecting a supply of fuel for the night, I came upon a heap of brush and leaves, and scraping off the top to see what was beneath, I discovered about half the carcass of a deer which apparently had been recently killed and partly eaten by a panther or Mexican lion, and the remainder cached in this heap for future use. Of course, under the circumstances, I had no scruples about appropriating the venison, and calling Brown and Holliday to my assistance we carried it to camp, where, after cutting off the ragged and torn portions of the meat, we soon had the balance spitted before a blazing fire. After making a hearty supper on our stolen venison, we raked a quantity of dry leaves close to the fire and turned into bed.

During the night, at various times, we heard the roaring of a Mexican lion (very probably the lawful owner of the larder that had supplied us with supper), and for fear he might be disposed to make a meal of one of us in place of venison, we took good care not to let our fire burn down too low. There is no animal, I believe, on the American continent, with the exception of the grizzly bear, that has ever been known to attack a man sleeping near a fire. The Mexican lion is, I think, described in books of natural history under the name of puma or South American lion. They are of a tawny or dun color, about the size of the East Indian tiger, have a large round head and a short mane upon the neck. Their nails are very long, sharp and crooked—coming to an edge on the inner side—as keen as that of a knife. Their roar is very similar to that of the African lion. They are fierce and strong, but cowardly; although when pressed by hunger, they have been known to attack men in open daylight.

One instance of this comes within my own knowledge. Several teamsters, with their wagons, were traveling the road from San Antonio to Victoria, and a teamster needing a staff for his ox whip, went to a thicket eighty or a hundred yards from the road to cut one; whilst occupied in cutting down a small sapling with his pocket knife, a Mexican lion stealthily crawled up behind him and sprang upon him before he was aware of its presence. The man's cries for help were heard by one of the teamsters, who hurried to his assistance. The only thing he had in the shape of a weapon was his ox whip, but with that he boldly attacked the lion, which, frightened by his approach and the loud popping of the whip, let go its prey and made a rapid retreat, but the poor fellow he had caught was dreadfully bitten and torn, and it was a long time before his wounds were healed. The Mexican lion is now rarely seen in Texas except among the dense chaparrals between the Nueces and Rio Grande Rivers.

As soon as it was fairly light we again started, and passing through a heavily timbered bottom, came to the Lavaca or Cow River, a small stream about thirty yards wide where we struck it. In going through the bottom we noticed several piles of rails and some clapboards, the first indications we had seen of settlements since we left Goliad. We also saw a drove of hogs in the bottom, which confirmed us in the opinion that there had been an American settlement somewhere in the vicinity. These hogs were of the genuine "razorback" species, and as wild and fleet as deer; consequently, although our hunger was almost as pressing as ever, we did not care to exhaust our strength in what we knew would be a hopeless attempt to capture one of them.

We swam the river without difficulty, and stopped an hour on the bank to rest ourselves and dry our clothes. We then went on, but as the bottom on that side was very wide, and the day was cloudy, we lost our way and it was nearly sunset when we reached the open prairie. A few hundred yards below where we came out of the timber we observed ten or a dozen horses "staked," and, on approaching them, we heard people talking in the woods nearby. I advised an immediate retreat from the locality, but for some reason Holliday came to the conclusion that the horses belonged to a company of Texan scouts, and proposed that we conceal ourselves in a clump of bushes from whence we could see anyone who might come to look after them and thus satisfy ourselves without running any risk as to whether the owners were Americans or Mexicans. Holliday's counsel prevailed, and Brown and I hid ourselves in a small bunch of bushes and Holliday in another.

A dog which was at the camp, all this time kept up an incessant barking, and probably it aroused the suspicions of the owners that someone was trying to steal their horses; at any rate, in a few moments after we had hidden ourselves, a strapping "ranchero" came out of the timber, and when he had looked to see if the horses had been disturbed in any way, he came as straight as he could walk to the bunch of bushes in which Brown and myself had taken our position and was just on the eve of entering it when he saw us. He instantly sprang back exclaiming, "Hey! Americanos! What are you doing here? Do you want to steal our horses?" He then made signs for us to follow him, which we did, knowing that resistance, weak as we were and without arms, would be useless, and that one shout from the ranchero would bring those in camp to his assistance. Holliday, as I have said, was concealed in a separate clump of bushes, and, as he kept quiet, the ranchero did not discover him.

Brown and I got up and followed the ranchero, but I was fully determined from the start not to follow him as far as his camp. I saw that his course would take him very close at one point to the timbered bottom, and as we went along Brown and I agreed upon a plan to escape from our captor, which was to follow him quietly until near the timber, and then suddenly break ranks and get under cover as speedily as possible. Then we were to take different directions and meet at the same place the next morning. The ranchero, although he could plainly see that Brown and I were unarmed, kept some paces ahead of us all the time, every now and then looking back to see if we were following. Before Brown and I separated I told him I would meet him at the Mexican camp the next morning, as it was probable they would leave it before we could return there.

In pursuance of our plan, as soon as we came very close to the edge of the timber, we suddenly left our ranchero without even saying *adios*, and in a moment we were hidden from his sight by dense undergrowth. When we so unceremoniously left our new acquaintance, we were so near the camp that we could plainly hear the rancheros conversing with each other, and the moment we made a break, our escort shouted to his companions to hasten to his assistance: "Here are Americans! Come quick and bring your guns!" But just at this juncture, Brown and I had some little matters of our own that required immediate despatch and we did not wait for the Mexicans to "come and bring their guns with them." Brown went one way and I another as soon as we entered the timber, and I never saw him again until several weeks afterwards when he came to the army on the Brazos.

The sun had just set when we entered the timber, and the night soon set in dark and cloudy. After going perhaps a mile, I concluded it would be impossible for the Mexicans to find me and I pitched my camp, which was speedily done by pitching myself on the ground at the foot of a tree on which there was a thick growth of Spanish moss, that served to protect me in a measure from a drizzling rain that commenced falling. I did not dare to start a fire for fear the light from it might bring the Mexicans to the locality, should they be in pursuit.

I had not felt so despondent since making my escape from Goliad as I did that night. My separation from my companions, my uncertainty as to their fate, the thought of my helpless situation, without arms of any kind to protect myself against the attacks of wild beasts and still more merciless Mexicans and Indians, together with the mournful howling of wolves in the distance, all conspired to fill my mind with gloomy forebodings and anticipations. However, notwithstanding such unpleasant thoughts and surroundings, I soon fell asleep and slept soundly until morning.

When I awoke day was beginning to break, birds were singing and squirrels chattering in the trees. The rain had ceased, and after brushing off the damp leaves that adhered to my clothes, my toilet was made, and I started back in the direction of the place where Brown and I had separated. I came out of the bottom very near the place where I had entered it the evening before, but no living thing was visible on the prairie as far as I could see, except some herds of deer and a flock of wild turkeys. I proceeded cautiously along the edge of the timber until I came to where the Mexicans had staked their horses. They were gone, and hearing no sounds from the woods in which they had camped, I ventured in to reconnoitre. Their fires were still burning, but the camp was deserted and there was nothing left to indicate the probable fate of my companions.

I was exceedingly hungry, and I searched the camp closely, hoping the Mexicans might have forgotten some remnant of their provisions when they went off, but I found no vestiges of eatables of any kind except a few egg shells. Leaving the camp, I returned to the prairie and traveled up and down the timber above and below it for several miles hoping to meet with one or the other of my companions. I continued my search for them until late in the evening, when having abandoned all hopes of finding them,

I struck out across the prairie in the direction I intended going Before I had gone a quarter of a mile I happened to look back towards the river and saw a house that had been hidden from my view, when searching for my companions, by a strip of timber. As I was suffering much from hunger, I concluded to return and make an examination of this house and premises, hoping I might find something to eat.

I approached the house cautiously for fear it might be occupied by a marauding party of Mexicans, but seeing nothing to excite my suspicions, I ventured up. Everything about the house—furniture broken and strewn in fragments over the floor, beds ripped open and their contents scattered around—plainly indicated that it had been visited by some plundering band of rancheros or Indians. However, in an outhouse near the main building, I found a piece of bacon and four or five ears of corn. The corn, I ground upon a steel mill that was fastened to a post in the yard, and starting a fire in one of the chimneys of the main building, I very soon prepared a substantial meal of "ask cake" and broiled bacon, to which I paid my sincere respects. By this time night had set in, and, spreading some tattered bed clothes left in the house upon the floor, I slept comfortably until morning.

The next morning, after making a hearty breakfast on ash cake and bacon, as there was no urgent necessity for hurrying forward, I concluded I would make another attempt to find my companions, and I again traveled for several miles above and below, along the edge of the timber, but seeing nothing of them I at length became satisfied that they had been captured by the Mexicans, or had gone on without waiting for me. The facts were, however, as I afterwards learned from both of them when I subsequently met them on the Brazos, about as follows: After Brown and I broke away from the ranchero and went off in different directions, he pursued Brown, came up with him and took him back to the camp. There they tied him securely to a tree, and then proceeded leisurely to cook and eat their supper. Brown, who could speak a little Spanish, told them he was starving and begged them to give him something to eat, but they said it was useless to do so as they intended to shoot him in the morning. He then told them if such was their intention to shoot him at once and not keep him tied up to a tree like a dog all night, but the Mexicans paid no attention to his request and when they had finished their supper, they laid down upon their blankets and went to sleep. Brown tried his best to untie himself, but the ranchero had fastened him so securely to the tree that he found it impossible to get loose, and was compelled to remain in a standing position all night.

The next morning, as soon as it was fairly light, one of the rancheros walked up to Brown and pinned a piece of white cloth to his breast, telling him it was a mark for them to shoot at. Four or five rancheros then stationed themselves a few paces in front of him, cocked their guns and presented them as if about to shoot. All this time, Brown, who had been rendered perfectly desperate by pain and hunger, was cursing the Mexicans as much as his imperfect knowledge of the language would permit. He told them they were a set of cowardly scoundrels, and that the bravest feat they had ever performed was the murder of unarmed and helpless prisoners, and so on. Brown said he was suffering and had suffered so excruciatingly from pain and hunger all night that he really wanted the Mexicans to shoot him and put him out of his misery, but they seemed much astonished at his boldness and sang froid, and the one in command of the party came to where he was tied, cut the ropes and told him to go, that he was "muy bravo" (very brave), and that in place of shooting him they would leave him to perish of hunger.

Then they saddled their horses and, mounting them, rode off. Some days afterwards Brown was again captured by a party of Mexicans, but in some way he managed to escape from them, and finally, more by good luck than anything else for he was a poor woodman, he made his way to the army on the Brazos.

Holliday, as I have before stated was not seen when the ranchero captured Brown and myself, and as soon as it was dark he left his hiding place and took his course across the prairie. Subsequently he had many narrow escapes from marauding parties of Mexicans and Indians. On one occasion a party of Mexicans pursued him so closely that he took refuge in a lake. He waded on until the water was up to his neck, when the Mexicans amused themselves for some time by firing off their scopets at his head, but fortunately for Holliday night came on and, under cover of the darkness, he skipped out and dodged his pursuers.

Another time, two runaway negro men caught him in a house to which he had gone in search of something to eat. They asked him if he was a Texan, and upon his replying in the affirmative they told him they intended to kill him. Whereupon they tied him securely in the room and went out, but in a few moments returned, each one with a heavy club in his hand, and they told him to say his prayers speedily, as they were going to beat out his brains. Holliday, however, "reasoned" the matter with them, telling them it wasn't fair to kill him for what other white men might have done to them—that he had never injured them in any way, etc. His talk seemed to produce some effect upon one of the negroes, but the other still insisted on killing him. Finally, however, the one who was inclined to favor him prevailed upon the other to abandon his intention of beating out his brains, and they said they would not kill him but would take him to the camp of some Mexican guerrillas near by. Holliday thought that this would be worse than "jumping out of the frying pan into the fire;" that such a proceeding would not be better than having his brains knocked out—and he urged all the arguments he could think of against it. At length, much to Holliday's relief, they agreed to let him go, and before they left they not only gave him provisions, but directions that enabled him to make his way through an unknown country to the Texan army under General Houston.

He came into Columbia, on the Brazos, about ten days after I did. Holliday was subsequently appointed to a captaincy in the Texas regular army, was again taken prisoner in the unfortunate Santa Fe expedition, carried to the City of Mexico, and, after his liberation, died of yellow fever on the voyage from Vera Cruz to New Orleans, and was buried at sea.

Giving up all hopes of finding my companions, I started out across the large prairie that extended in the direction I was going as far as my eye could reach. The game on this prairie was more abundant than I had seen it elsewhere. I am sure that frequently there were a thousand deer in sight at a time. Here, too, I first saw the pinnated grouse, or prairie hen. At first I supposed the call of the cock was the distant lowing of wild cattle, some of which were grazing on the prairie. Wild turkeys were also numerous, and so unused to the sight of man, that they permitted me at times to approach within a few paces of them.

During the day I saw several parties of Mexicans or Indians on horses, but they did not come near me. About three o'clock in the evening I reached the timber on the Navidad, where I stopped to rest a while and lunch on some of the ash cake and bacon I had brought along with me. I then proceeded on my course through the bottom, and after going probably half a mile I came to the Navidad river, at that place thirty or forty yards wide. It was swollen by recent rains and not fordable, so I was compelled to swim it, which I did easily, stripping off my clothes and tying them on a piece of dry wood, and pushing it before me as I swam.

As soon as I reached the bank I dressed myself and continued my course through the bottom, which was much wider on that side. I had gone perhaps half a mile, when my attention was drawn to the continuous barking of a dog in the direction from which I had come. At first I did not notice it particularly, supposing it was some dog left behind by the settlers on the Navidad when they fled from the invading Mexican army. But at length I observed that although I was traveling at a pretty rapid walk, the barking of the dog seemed to be nearer and nearer to me, and I suspected he was trailing me and that probably there was someone with him. I therefore hurried on as fast as possible, and in an hour or so came to the open prairie on the north side of the river. All this time I could hear the baying of the dog at apparently about the same distance behind me as when I first noticed it. I was sure then he was trailing me, and never halted for a moment, but continued on my course into the prairie for several hundred yards and then turned short round and retraced my steps to the edge of the timber. I sprang as far as I could to one side and went down the edge of the timber about a hundred yards to a fallen tree, among the limbs of which I concealed myself, and from whence I could have a distinct view of anything coming out of the bottom at the point I left it.

After I had thus holed myself, the barking of the dog grew louder and nearer every moment, and in a little while I saw the dog, followed by three Indians, emerge from the timber, precisely at the point where I had left it. One of the Indians held the dog by a leash and was armed with a gun, the other two had their bows and lances. If I had been armed with the poorest pot-metal, muzzle-loading shotgun that was ever manufactured at Birmingham, I would not have feared them, but as I had no weapon more formidable than the scissor blade given me by Brown, I laid low and watched them from my hiding place. When the Indians following the dog came to the place in the prairie from whence I had turned back on my trail, the dog lost it of course, but the Indians (taking it for granted, I suppose, that I had gone in the same direction) urged and led the dog that way until finally they went out of sight. If I had not thrown them off my trail in the manner described, there is no doubt I would have lost my scalp on that occasion, and I took considerable credit to myself for having beaten them at their own game.

I remained but a little while in the hiding place after the Indians left. But the course I wished to travel was the one they had taken, and for that reason, and because my provisions were nearly exhausted, I determined to keep up along the edge of the timber, hoping to find some settlement and replenish my larder. I followed up the margin of the timber for several miles, and at length came to a clearing, on the opposite side of which I saw a house. I cautiously advanced towards the house until I was satisfied it was not occupied, and that I could venture up with safety. On entering it I found that a marauding party of Mexicans had lately been there and appropriated to their own use whatever there might have been eatable on the premises. I searched the house thoroughly, but could find nothing in the way of provender.

By the time I had finished my fruitless search for something to eat, the sun was about setting and, as there was a bed in the house, which looked very inviting to me after sleeping so long on the ground, I concluded to accept the invitation and pass the night in it. After a very frugal and unsatisfying repast upon the small remnant of ash cake and bacon in my knapsack, I turned into my bed and was soon fast asleep.

It must have been near midnight when I was aroused by some noise. I listened attentively and soon ascertained that the noise was nothing but the grunting of several hogs that had taken up their quarters under the house whilst I was asleep. The house was set upon blocks, a foot or so above the ground and the space beneath the floor was therefore sufficiently roomy for their accommodation. The floor was made of puncheons or slabs, which were held in their places solely by their weight. Hunger as well as necessity is the mother of invention, and it occurred to me that I might bag one of these porkers by quietly lifting a puncheon immediately above the spot where they were lying and then quickly grabbing the first one I could get hold of.

I therefore got up from my comfortable bed, and after listening awhile to their grunting so as to ascertain what part of the floor they were under, I slowly and noiselessly lifted a slab above them and laid it aside. Thrusting my arm down through the opening I had made, I felt around until my hand came in contact with the leg of a hog, when I suddenly seized it, and the row began. I had got hold of a hog much too large for me to manage well, and found it no easy matter to induce him to come up into my comfortable quarters. He struggled vigorously to get loose, squealing all the while in the most ear-piercing manner, and for some time I thought it very doubtful how the contest would end—whether

I would succeed in hauling the hog up into the room, or the hog in dragging me under the floor. But I knew if I let go there would be no pork steaks for breakfast, as the other hogs had been frightened by the squealing and struggling, and had left for parts unknown. But the idea of having no steak for breakfast gave me more than my usual strength, and at last, but not until he had cut me severely with his hard hoofs and rasped a good deal of the skin off my knuckles against the sharp edges of the puncheons, I drew him by main "strength and brutality" into the room and replaced the puncheon.

I had secured my hog, but how to kill and butcher him was the next question. I had nothing to do it with except one of the blades of the little pair of scissors given me by Brown, and that I knew was totally inadequate for the purpose. I could find nothing in the room that would do, so I slipped out, carefully fastening the door after me, to see if there was anything about the premises with which I could dispatch the porker. The moon was shining brightly, and I looked all around for something that would enable me to convert my hog into pork, but could find nothing better than a large maul that had been used for splitting rails, and with this I reentered the room and made a determined assault upon the hog. The maul, however, was so heavy and unwieldy I could not handle it with sufficient celerity to inflict a stunning blow. Round and round the room we went for a quarter of an hour or more, the hog squealing all the while and his hoofs clattering and rattling on the puncheons and making altogether such a racket as might have been heard at the distance of half a mile. At last, however, I got a fair lick at his cranium, which brought him to the floor, where I finished him by continuous mauling.

When the bloody deed had been committed, I was so completely exhausted that I tumbled back on the bed, was asleep in a few moments, and did not awake until the sun was high in the heavens. I got up, and the first thing I did was to drag my hog to a spring near the house, where I butchered him after a fashion, with a piece of broken drawing knife I picked up in the yard.

After finishing this job I started a fire, and roasted four or five pounds of the pork for breakfast.

When I had breakfasted, I packed as much of the pork as I could carry in my knapsack, and started up the bottom again, keeping close to the edge of the timber so that I might readily take shelter in the event that I should meet with a party of Mexicans or Indians. I had come to the conclusion by this time that previously I had been steering my course too low down the country, and I thought it best to keep up the river some distance before I resumed it again, in order to avoid the lagoons and swamps which I supposed abounded in the vicinity of the coast.

I traveled five or six miles without seeing anything worthy of note, and at noon stopped an hour or so at a pool of water to rest and cook some of my pork, and to barbecue the remainder so as to prevent it from spoiling. It was late in the evening before I started again, and about sunset, not finding another house, I concluded to encamp in a point of timber near a pool of water.

Just after I had turned into a bed of dry grass for the night, I saw a light spring up, apparently five or six hundred yards above, on the edge of the bottom, and I concluded to get up and see what caused it. The moon had not as yet made her appearance, and I thought I could reconnoiter the locality with safety, even if the light should prove to be from the campfire of Mexicans or Indians. Guided by the light, which continued to shine steadily, I went perhaps a quarter of a mile, when I saw that it came from the chinks of a small log cabin. I approached it silently, and when near it, I saw there were several other cabins near it, but no lights were visible in them. The chinks between the logs of the cabin in which the light was shining were all open, and I carefully crept to the side nearest me and peeped through one of them.

I had heard for some time a queer kind of rasping sound proceeding from with the cabin, for which I could not account until I looked through the chink, and then I saw a Mexican soldier sitting on the floor, shelling corn into a tub, which he did by rasping the ears on the edge. He had on his shot pouch and powder horn, but his gun, I noticed was leaning against the wall next to me, and as there was an opening between two of the logs it was leaning against wide enough to shove my arm through, it occurred to me that possibly I might be able to draw the gun through this opening before the Mexican was aware that anyone was in the vicinity, as his back was turned towards me. So I reached in, seized the gun cautiously, near the muzzle, and began to draw it slowly through the chink between the logs. There is no doubt I would have succeeded in my attempt to get the gun, but when the barrel was fairly outside and I felt sure I had secured the prize, to my great disappointment the breech was so large that it stuck hard and fast between the logs. In my effort to pull the gun through, I unavoidably made some noise that attracted the attention of the soldier, and he turned and uttered an exclamation of fear and astonishment when he saw his gun thus mysteriously disappearing through the chink in the cabin, and he instantly sprang forward and clutched it by the breech.

The noise aroused three or four dogs sleeping near the cabin, and they began to bay me furiously. I was sure there were more Mexican soldiers in the adjoining houses, and thinking I might find a healthier location than the one where I was, I made off at double quick for the bottom, closely pursued by the dogs. When I reached the timber, I picked up a club, turned upon the dogs and drove them back. I heard a good deal of shouting and "carajoing" at the cabins, but as the night was quite dark I had no fear of being pursued, and leisurely took my way along the edge of the timber. When I had got I suppose a mile from the cabins, I went into the timber and encamped in a secure place.

My failure to get the soldier's gun was a great disappointment to me. Every house I had visited since I struck the settlements, I had searched closely for a gun, hoping that one might have been left by the occupants when they hurriedly fled before the invading army, but my search was always fruitless. People had abandoned a great deal of valuable property, but whatever arms they had they carried off. I had an abundance of ammunition, for at one of the houses I had searched I found powder and shot, which I secured, and all I lacked was a gun. I would willingly have given all the money I had in the world (amounting to seventy-five cents in specie) for the poorest pot-metal gun that was ever manufactured, and taken the chances of its bursting whenever I fired it.

Just at daylight I was aroused from my slumbers by the clucking and gobbling of wild turkeys. I had encamped very near a large roost, and as I made no fire I had not disturbed them. Many of the trees in the vicinity were literally filled with them, and they were so tame I could easily have killed one with a bow and arrow if I had had them, and I determined I would try my hand at manufacturing these primitive weapons, if I could find some suitable tool to work with.

After I had reconnoitered from the edge of the timber and ascertained that there were no Mexicans in sight, I went on up the bottom three or four miles, and then struck across the prairie in the direction I had been traveling. My route was through an open prairie interspersed with mots or groves of timber. In one of these I stopped about noon, and broiled a piece of my pork for dinner. After resting an hour or so I continued on my way, and about sunset came to some timber bordering a small stream. I had scarcely entered this timber, which was open and free from undergrowth, when I noticed several large wolves trotting along behind me. Every now and then they set up a howl, which was answered by others in the distance, and before long numbers of them had gathered around me, attracted, I suppose, by the howling of those I had first seen, or by the smell of the fresh meat I had with me. I had no fear of an immediate attack from them, nevertheless, I hurried on as fast as I could until I came to the small stream I have mentioned, on the bank of which I pitched camp, near a large fallen tree that would afford sufficient fuel to keep a fire burning all night.

I am confident if I had not had a fire that night, the wolves would have torn me to pieces; as it was, they sometimes ventured up to within a few feet of the fire, howling and snarling, and evidently inclined to make a dash at me at all hazards. It was impossible to sleep, so I took my spite out of them by occasionally throwing a fire brand amongst the crowd. This would silence them for a moment, but they would soon begin their howlings again. Towards daylight they raised the siege and departed, and I got a little nap before sunrise.

Today, while crossing another large prairie, I saw in the distance a considerable body of Mexicans or Indians, I could not tell which, who were traveling at a rapid rate, and I soon lost sight of them. In this prairie I passed many herds of deer, generally fifty to a hundred in a herd, which were so gentle they frequently permitted me to approach within a few paces of them before they noticed me at all. I also saw several droves of mustangs, which were much wilder than the deer, and invariably whenever I got within five or six hundred yards of them they would raise their heads, gaze at me for a few moments, and then with much snorting and cavorting they would go off like the wind, and never slacken their speed as long as they were in sight.

In a small grove of timber where I had halted to rest awhile, I saw for the first time a horned frog. I had heard of the tarantula and centipede of Texas, and supposing the harmless frog was one or the other I picked up a stick about ten feet long (not venturing to approach nearer such a poisonous reptile) and mashed him as flat as a pancake.

Continuing my course, at sunset I came to a belt of timber bordering another small stream. On the bank of this stream was an Indian encampment that appeared to have been occupied a day or so previously. Several of their fires were still smoking, and from their number I supposed there were thirty or forty in the party. Around these fires was scattered a great quantity of bones, mostly those of deer, though the head of a mustang here and there showed that they varied their diet by an occasional feast on horse flesh.

A cold misting rain had begun to fall just before I came to this camp, and seeing it was likely to continue through the night, I took possession of a shanty built of small poles and covered with slips of bark. In this I stowed myself and baggage and made myself perfectly at home. With a large fire in front of it and plenty of hog, but no hominy, I passed a very comfortable night, serenaded as usual by wolves.

Next morning the rain had ceased, and the sun was shining brightly when I woke up. Cooking a piece of my pork, I made a hasty breakfast for fear the owner of the shanty might return and ask me to pay for my night's lodging, and again started on my journey.

During the day I saw several signal smokes, made I suppose by Indians, but they were a long way off. These signal smokes are curious things. Often when traveling over the plains of Western Texas, I have seen a column of smoke rise perpendicularly into the air (no matter how strong the breeze might be blowing) to a great height, when it would spread out at the top like an umbrella, and after remaining stationary for a moment, "puff," it would suddenly disappear, to be answered perhaps by another, twenty or thirty miles away. They are no doubt intended for signals to warn others of the proximity of foes, and to indicate their own position. I have asked many old frontiersman how it was the Indians made smokes, but none of them could ever explain the matter satisfactorily to me. I have occasionally seen four or five of these signal smokes rising up in various directions at the same time.

Today, for the first time, I saw what I know now was a tarantula, a very large and exceedingly venomous spider, that haunts the dry and elevated prairies of Western Texas. They are not often seen in the timbered lands or in the immediate vicinity of settlements. The body of a full grown one is as large as a hen's egg, and is covered with scattering hairs or bristles. They have two curved fangs protruding from the mouth, about as long and very similar in appearance to those of the rattlesnake. When provoked they are very pugnacious, rising upon their hind legs and springing towards the assailant five or six inches at a time in successive leaps. The Mexicans say their bite is certain death, and one can readily credit the assertion after seeing them.

I made but little if any progress today, for not long after I had started it clouded up and commenced misting again, so that I lost sight of the timber towards which I was steering my course. Finally I became completely bewildered and after wandering about all day I came to a belt of timber I had good reason to suppose was the same I had started from in the morning. At any rate the sun just then showed itself for a few moments, and I found I was traveling in the direction directly opposite the one I should have pursued.

It was too late to take the prairie again, and I picked out a suitable place for camp, started a fire and cooked some of my pork for supper, which for want of salt was getting to be rather too much tainted to suit the taste of anyone but a Frenchman. During the night the wolves favored me with another concert of howlings, but they were much less impudent than upon a former occasion, and did not approach near enough to enable me to salute them with fire brands.

In the morning I rose early and, unpacking all the pork I had left, I spitted it on sticks stuck up before a blazing fire I thought by roasting it in this way to keep it from spoiling entirely. The clouds had blown off and the sun shone out warm and pleasant, and having eaten some of my roasted pork which had decidedly too much of the "gout," I started out again across the open prairie.

This time I made the trip without difficulty, and about mid-day I came to a small stream which I afterwards learned was called the Tres Palacios or Three Palaces. How it acquired the name I cannot say, but I am sure I saw no palaces in its vicinity. Where I crossed it, I noticed a few small cedar trees growing near the bank, and I determined to cut one of them down and make a bow. This was no small job, as you may suppose, considering I had nothing to cut it with except a small piece of the blade of a drawing knife—the same I had found at the house where I killed the hog, and which I had carried in my knapsack ever since. By the time I cut the sappling down, I was both tired and hungry, so I knocked off work to rest a while and cook some pork. I then resumed my task and, chopping off about six feet from the butt end of the sapling I split it into four pieces with a wooden wedge and maul. From these I selected the one that was freest from knots and other defects, out of which, by patience and perseverance and with the aid of my piece of drawing knife, I manufactured a very good bow. Arrows I knew I could easily get anywhere in the bottoms among the thickets of swamp dogwood or young cane.

By the time my bow was finished night came on, and I pitched my camp near the creek in a little open space completely surrounded by a thick growth of underwood. Here I built my fire, warmed over some of my roasted pork and, after supper, turned in to a bed of Spanish moss which I had gathered from a tree nearby.

The next morning I gave the finishing touches to my bow and then for the first time it occurred to me that I had nothing that would answer for a string. I tried to make one of the bark of several shrubs, and of the leaves of bear grass, but although I taxed my ingenuity to the utmost, I failed to make a cord strong enough for the bow, and I had at last to abandon the attempt altogether.

This was a great disappointment to me as I had calculated largely on supplying myself with an abundance of small game by means of my bow. I had heard of people having "two strings to their bows," and yet under the most pressing necessity I was unable to get one for mine—which convinces me that things are very unequally divided in this world.

The day was so far gone when I had finished my unsuccessful attempt at cord making, that I thought it best to remain where I was for the night and make a fresh start in the morning. It must have been twelve or one o'clock, when something awoke me, and finding that my fire had pretty well gone out, I was just in the act of getting up to throw some sticks on it, when I heard the stealthy but heavy tread of some large animal nearby. I laid still and listened attentively, and was convinced there was some heavy animal cautiously approaching the spot where I was lying. Just then, fortunately probably for me, a chunk rolled off a log I had placed behind the fire, and blazed up brightly. By the light thus made, I saw distinctly either a large panther or Mexican lion, not twenty feet distant, crouching down as if about to spring upon me. I instantly jumped, and seizing my bed clothes (the dry Spanish moss I had gathered) I threw it on the fire and it blazed up at once as high as my head. This must have frightened the animal, whatever it was, for when I turned to look it was gone. Possibly it did not intend to attack me, but the way in which it had approached me was, to say the least of it, very suspicious. The loss of my "bed clothes" did not discommode me much, as I sat up the balance of the night to keep my fire supplied with fresh fuel, although the night was quite warm.

As soon as the sun rose, I made haste to leave the locality where I had passed such an unpleasant night. Late in the evening I came to an extensive body of timber, in which I supposed I would find a considerable stream. On the edge of this timber I saw a house, and, as by this time what remained of my pork was so strong of the "gout" that I don't think even a Frenchman would have relished it, I determined to go to the house and search for something to eat. I entered the woods some distance below it, and kept under cover until I was near enough to see there was no one about, when I ventured up. On entering I soon saw that it had been ransacked by the Mexicans, who had consumed or taken away whatever there might have been in it in the way of eatables. In the vicinity, however, as I was leaving, I came across a half-grown hog, which evidently had very recently been shot by someone, who had taken only a small part of it, and I appropriated as much of what was left as I could conveniently carry. As the sun was about setting, I went some distance into the timber, so that the light from my fire would not be visible to anyone passing along the prairie, where I "bivouacked" for the night at the foot of a tree.

By sunrise I was up and on my way again, crossing in a mile or so a considerable creek. Today I passed over a country of mostly prairie, but interspersed here and there with groves of live oaks, hackberry, etc., which gave it a park-like appearance. In one of these groves, thickly settled with underbrush, I stopped to rest, and was just in the act of leaving it, when I heard the tramping of horses' hoofs and the jangling of spurs and other accoutrements. Looking through the bushes I saw about twenty Indians slowly jogging along in single file upon their horses. They had no guns and were armed only with bows and lances. They rode within thirty paces of where I was lying low, but did not halt, and in a few moments they were hidden from my view by another grove. I remained where I was half an hour longer than I would have done otherwise, in order to give these Indians full time to get out of my way, and then proceeded on my course. A little before sunset I came to a clear running creek, on the farther side of which I encamped. (At that time, all the creeks and small water courses, and even the ponds in Western Texas were clear and pure, but now many of them have lost that character to a greater or less extent, owing to the cultivation of adjacent lands and the tramping of stock.)

I had made my camp beneath some low spreading live oaks, which appeared to be a favorite roosting place for wild turkeys. Just at dusk they came flocking into them from every direction, and they were so unused to being hunted, I could easily have killed one with a pocket pistol—but as I didn't have the pistol I had to content myself with roast pork instead of roast turkey.

I had noticed before dark that a very extensive prairie lay to the north and east, and I was up and on my way the next morning before daylight, in order that I might reach the timber on the opposite side as speedily as possible. I ran but little risk comparatively when traveling in timber, but on the open prairie I was in constant danger of being picked up by parties of Mexicans or Indians. I pushed on as fast as I could until noon, when I stopped to rest in a grove near a small lagoon that seemed to be well stocked with fish, for I saw numbers of bass and perch swimming in the shallow water near shore. On the margin of this lake I found some wild onions growing, which I dug up and ate raw, and which were a great treat to me, as I had not had anything in the vegetable line, fresh and green, for a long time.

In the evening I continued on my way across the prairie on the farther side of which I could see a long line of unbroken timber stretching from northeast to southwest, as far as my eye could reach. It was nearly night when I came to this timber, and I had gone but a little way in it, when I saw a large river before me, which I knew must be the Colorado. The river was very high and rapid, and I thought it best to encamp for the night and wait until morning before I attempted to swim it. Where I struck it, it was about two hundred yards wide and much swollen by recent heavy rains, and although I was a good swimmer, I felt some hesitation the next morning in "taking water." However, I looked around and found a suitable piece of dead timber, to which I tied my boots and clothes, and launched forth with it on the turbulent stream, pushing it before me as I swam. Finally I made a landing safely on the north bank of the river, but was carried by the strength of the current a considerable distance below the point where I had entered the water.

After resting myself awhile and drying my clothes, I took up the line of march again through a heavily timbered bottom about a mile and a half wide, from which I at length emerged into the open prairie. Without halting I continued on my course until late in the evening, when I came to the timber on old Caney Creek. Along this creek, which apparently in times gone by was the bed of the Colorado River, from its head to its mouth, a distance of sixty or seventy miles, there was a continuous cane brake. Where I struck the timber on old Caney, there had been a considerable settlement, as some four or five houses were in sight, but on examination, I found that all of them had been plundered by Mexicans, who had taken everything of any value left on the premises.

At one of these houses whilst searching the rooms to see if anything in the way of provisions had been overlooked by the Mexicans, I heard a hen squawking as if some "varmint" was in pursuit of her. I stepped to the door to look out, and saw a hen racing around the yard and a very large wildcat following her closely. Having seen nothing eatable anywhere, except this hen, I determined to put in a bid for her myself, and picking up a billet of wood, I stepped out boldly towards the cat. When he saw me coming, he quit his pursuit of the hen, but showed not the slightest disposition to abandon the field. I advanced to within a few paces of where he stood humping his back and showing his teeth, and threw the stick I had in my hand at his head. I missed my aim, but struck him a severe blow on the side, and instantly he gave a scream and sprang furiously towards me. I retreated precipitately and ingloriously for the house, which I reached just in time to rush into the door and slam it to in the face of the infuriated cat. If I had had a few feet further to go, he would have nabbed me to a certainty. The cat stopped some time in front of the door, as if he intended to besiege me in the house, or was bantering me to come out and give him a fair fight, which, under the circumstances, I declined doing, but after a while he went off leisurely towards the woods and I saw him no more.

In the meantime the "bone of contention," the hen, had gone to roost in tree nearby. She undoubtedly owed her life to me, but for a very little while, for after dark I climbed up to her roost, grabbed her by the leg, and wrung her neck. With my prize, I retreated as speedily as possible to the house, for fear the wildcat might return and assert his claim to it again, and as I had no weapon I was very sure be would get the better of the contest and the hen too.

I remained all night at this house, and after breakfasting on the hen I had *saved* from the wildcat, I started off down the bottom to reconnoitre the country in that direction. When I had gone a mile or two I came to a small prairie connected with the main one by a very narrow neck and surrounded everywhere else by thick woods and cane brakes. This I concluded to explore, and after proceeding some distance in it, I saw there was a house at the farther end. When I had approached within a hundred yards of the house, a half a dozen dogs came rushing out of it, seemingly with the intention of tearing me to pieces. I picked up a stick to defend myself, but when the dogs got near enough to see that I was an American, instead of attacking me they began to leap and jump around me as dogs do when they see their masters after a long absence. How they found out so quickly I was an American, I do not know, for exposure to sun and weather had tanned my complexion, until it was as dark as that of a Mexican or Indian.

With my escort of dogs I went to the house, and entering it, saw at once that the Mexicans had not been there, for everything remained, evidently, just as it had been left by the occupants-furniture untouched, cases filled with books and articles of wearing apparel, cribs with corn and a smokehouse containing at least a thousand pounds of bacon. In a kind of shed room I also found a barrel of brown sugar and half a sack of coffee, and in the crib, besides corn, a quantity of potatoes and pumpkins. There were a great many chickens and ducks in the yard, which no doubt, had been protected from "varmints" by the pack of dogs that still continued to escort me about the premises. In the smokehouse as I have said, there was a large quantity of bacon, and the first thing I did was to take a middling and cut it up for the dogs. I then built a fire in one of the chimneys and in a little while had cooked for myself a first rate dinner together with a cup of coffee, the first I had tasted since leaving Goliad. After dinner I turned in to one of the beds in the house and had a comfortable snooze.

When I awoke I got up and continued my investigations. In a back room I found quite a library, a rare thing at that time in Texas. I found also many articles of clothing in a closet, some of which fitted me tolerably well, and from which without any fear of being arrested for "petit larceny," I replenished my scanty wardrobe. Among other things I found in this house—something I wished for exceedingly—was a gun, but unfortunately it was without a lock, and consequently useless. Not far from the main building there was a row of log cabins, that evidently had served as negro quarters, which induced me to believe that the place belonged to some well-to-do cotton planter.

As I had been much weakened by starvation and fatigue and the exposure I had undergone in my route through the wilderness, I concluded I would stop over a day or two at this house and recuperate my strength a little before I set out on my journey again. There were beds in several of the rooms, in one of which I slept at night, while my pack of dogs kept watch outside. These dogs were not mongrels or "curs of low degree," neither were they of the "suck egg" breed, as was evident from the fact that although they were in a starving condition when I came, and that the chickens had laid their eggs almost everywhere in the house and yard, not one had been touched by them—for which I was thankful, being particularly fond of eggs myself.

I remained several days in my comfortable quarters, feasting on the good things I found in them, and reading books I selected from the library. On the evening of the third day of my sojourn at the house, feeling a little unwell (I rather think I had been indulging somewhat too freely in fried chicken), I concluded I would take a short stroll around my domains by way of exercise. After going a few hundred yards I turned to take a bird's eye view of my surroundings, and I exclaimed as Crusoe did on his island:

I am monarch of all I survey,
My right there is none to dispute,

except, I mentally added, a marauding party of Mexicans or Indians, and now and then a wildcat.

Whilst passing through some tall grass, I came very near treading on a rattlesnake, the first I had seen in Texas, although some portions of the country I had passed over were much infested with them; but the season then was hardly far enough advanced to bring them out of the dens or holes in which they take up their winter quarters. Often since, when passing over some of the uninhabited plains between the Nueces and Rio Grande rivers, I have found them so numerous in particular localities, that I was scarcely ever out of hearing of the sound of their rattles. They are not, however, nearly so vicious in Texas as they are in some other countries, and seldom attempt to strike, unless attacked. I have slept with them, ridden and walked over them frequently, and instead of trying to bite me they always did their best to get out of the way—except on one occasion.

I was stalking some deer one day on the prairie, when I stepped upon a rattlesnake lying coiled up in the grass. I knew even before I saw it, by the peculiar soft squirmy feel under my foot that I had put it on a snake, and I promptly lit out without waiting for orders. As I did not wish to shoot him for fear of alarming the deer, and as they are easily stunned by a very slight tap on the head, I drew the ramrod from my rifle and gave his head a smart blow with it. I then mashed his head by repeated blows with the breech of my gun and thinking, of course, I had killed him, I went on after the deer. Two days subsequently, when passing the place again, that same snake came very near biting me. I knew it was the same, for one of his eyes was out, and his whole head bruised and bloody from the blows I had given it with the breech of my rifle. I really believe be recognized me as the author of all his ills, for when I attempted to go near him he would raise his head a foot or more from the ground and, with his rattles going incessantly, would glare at me with his one eye in the most vindictive way.

I determined to make sure of him this time and, leveling my rifle at his head, I took good aim and fired. The bullet knocked his head into fragments, and one of the pieces struck me on the forehead, making a slight wound. The idea immediately occurred to me that I had been struck by one of his fangs, and that I was fated to be killed by this particular snake. However, after bathing the scratch in a pool of water, and finding that my head had not swelled up as big as a bushel, I went on my way, congratulating myself upon my second escape from my vindictive foe.

But to return from this digression, to my story. On the morning of the fourth day of my sojourn at this house, I concluded I had regained my strength sufficiently to take the road once more, or rather the woods and prairies. Preparatory to leaving, I packed up as much sugar, coffee and bacon as I could carry, together with five or six pounds of meal, which I had ground upon a steel mill. I also put a tin cup in my knapsack, and several other articles which I thought would be useful to me. When ready to start I stuck a couple of carving knives (which I had also found at this house) in my belt and, bidding adieu to my dogs, after I had given them middlings enough to last them for a month, I set out on my travels again. But, to my great dismay, when I had got a few hundred yards from the house, I found I had not consulted the wishes of the dogs about leaving them, and that the whole pack was following close at my heels—suspecting, I suppose, from the preparations they had seen me making, that I was going for good.

I tried to drive them back by throwing sticks and other things at them, but it was all to no purpose. They would stop whenever I did, but the minute I started they followed on. I knew it would be impossible for me to travel safely through a country in which I would be liable at any time to meet marauding parties of Mexicans and Indians with a half dozen dogs at my heels, and finding I could not get rid of them, I determined to go back to the house, wait there until night, and then quietly leave them. So I returned, and passed another day very pleasantly at my house, looking over the books in my library, and cooking and eating at short intervals.

Before I retired to my apartment, I noticed particularly where the dogs were sleeping, and about midnight I got up, quietly shouldered my pack of provisions, and left the house. I had gone perhaps half a mile down the edge of the cane brake when I heard the pattering of feet behind me, and in a few moments one of the dogs came up. I beat him severely with a stick, but he only whined and crouched down at my feet. Finally, I determined to kill him with one of my butcher knives, but as I grasped him by the neck, and drew my carving knife, he looked up at me so piteously that I hadn't the heart to use it, and abandoned my murderous intention.

I thought I could manage to keep one dog under control, and that the risk I ran of being killed or captured would not be increased to any great extent by having a dog with me; besides, I came to the conclusion that the company of a dog was better than none. Like the Frenchman, I think that solitude is very pleasant at times, provided there is someone with you to whom you can say "how delightful is solitude."

The dog that followed me was a very large and powerful one—a cross, I think, between the English bull and the Newfoundland. I found him to be tractable and, at the same time, as courageous as a lion. In a few days I had him perfectly under control, could make him lie down at a word and remain at camp to guard it when I went off foraging or reconnoitering. I named him Scout.

After traveling a mile or two down the brake, I thought I had gone far enough to get away from the other dogs, and I encamped for the balance of the night near a lagoon. I heard no wolves at this camp, but several times during the night I was roused by the noise made by some large animal forcing its way through the cane. I suppose it was a bear, as I noticed next morning a great many tracks in the soft ooze near the margin of the lagoon.

Whilst lying awake the next morning, upon my bed of dry leaves, my attention was drawn to a rustling among them, and turning them over, I found an ugly reptile about six inches long, which I thought then, and know now, was a centipede. Not fancying such a bedfellow, I quickly dispatched him with a stick. They resemble somewhat the reptile called the "thousand leg worm," but they are much larger and flatter, and although they are well provided with legs, they have not quite a thousand. They are of a dark brown color on the back, and the underside a dirty white. Their tail is forked, and has a long sting in the end of each prong, besides smaller stings on each foot and, to complete their means of inflicting wounds, the mouth is furnished with fangs. They are a disgusting looking varmint, and are said to be very venomous. An old Texan speaking about them, said: "When they wound you with their feet alone, it hurts considerable; when they sting you with their forked tail it's a great deal worse, but when they pop you with all their stings and bite you too—say your prayers."

As soon as I had cooked and eaten breakfast and Scout had cleaned the dishes by licking them, I began to search again for a road that would lead me across the brake. Failing to find one after searching for several hours along the edge of the brake, I determined, if possible, to cut my way through it. I therefore attacked the cane, green briers and bushes with a carving knife, and after working faithfully till late in the day, I found I had gone about three hundred yards. Such slow progress was exceedingly discouraging, for at that rate, if the brake was as wide as I thought it to be, I would be several weeks getting through it. There were a few scattering trees among the cane, and in order that I might be able to form some idea of the width of the brake, I climbed one of the tallest, from whence I could see an ocean of cane, extending at least four miles in the direction I wished to go, and beyond the scope of vision to the northwest and southeast. The length of time and the amount of labor that I knew would necessarily be required to cut my way for so long a distance through this dense mass of vegetation, induced me to give over the attempt, and, descending from the tree, I took the path I had cut back to the prairie.

Feeling considerably fatigued by my labors, when I got to the edge of the brake, I sat down at the root of a large tree to rest awhile. Gradually I fell into a doze, from which I was suddenly aroused by the growling of Scout, and a scuffling, scratching noise overhead. Looking up, I caught a glimpse of some huge black animal sliding down the tree a few feet above my head. I sprang off quickly to one side, and at the same instant a bear struck the ground and took his way into the cane, which popped and cracked as if a wagon was going through it. It would be hard to say which was the most frightened, I or the bear, and even Scout was so demoralized by his unexpected appearance that he made no attempt to pursue him. The bear, of course, was up the tree when I took my seat at the foot of it and, as the tree was densely covered with Spanish moss, I had not noticed him. From my protracted stay at the foot of the tree, I suppose the bruin had come to the conclusion that I was laying siege to him regularly, and getting desperate, he had charged down upon me in the manner I have related. Had I known it was a bear when I first caught a glimpse of him, I should not have been alarmed, as I had never heard tell of their attacking anyone except when wounded and brought to bay.

Several days afterwards, however, two of them exhibited such evident signs of hostile intentions towards me that I was induced to believe that they were not so non-combative as generally supposed.

After this little adventure, I continued on along the edge of the brake, hoping I might find some road or trail leading across it. I examined every nook and indentation, and finally came to quite a large trail leading from the open prairie towards the brake. Along this trail the old traces of wagon wheels were distinctly visible. I followed it for some distance running almost parallel with the brake, and at length came to where it abruptly turned and entered it. After crossing a strip of cane about two hundred yards wide, with a small lagoon near the center of it, spanned by a rude bridge of logs, I came to a small prairie perhaps a mile in length and half a mile wide, a considerable part of which had been in cultivation. At the farther end of this prairie I saw a house, to which the trail I was following seemed to lead.

When I had approached to within three or four hundred yards of the house, I halted for a few moments to make sure whether or not there was anyone about the premises. I heard the crowing of chicken cocks and the squealing of pigs, but as I saw no smoke issuing from any of the chimneys or any other signs to indicate that the house was occupied, I ventured up. There were a great many chickens, ducks and pigs in the yard, but no dogs came to welcome us.

The house was a comfortable log building, consisting of four rooms with a wide passage between them and a broad piazza in front, and was sheltered by some large live oak and pecan trees. Everything in the house remained just as it was when abandoned by the occupants, which convinced me that it never had been discovered by the Mexicans. Indeed so secluded was the locality and so completely hidden from view by the strip of tall cane on the lagoon before mentioned, that no one passing along the main prairie outside would have suspected there was a settlement in the vicinity. This house was furnished even in better style than the one I stopped at last, which, together with the number of outhouses and negro quarters, convinced me it had been the residence of a wealthy planter. In the barns and cribs I found a large quantity of corn, potatoes, etc., and plenty of sugar and coffee in a storeroom.

By the time I had made a thorough examination of the premises, the day was pretty well spent and I determined to take up my quarters for the night in the house. Besides, it had clouded up and a cold, misting rain had begun to fall. I therefore proceeded to make myself at home without the least ceremony. I lolled upon the sofa, read the books, smoked a pipe (which the proprietor of the premises had left behind in the hurry of departure, with a box of tobacco), and after I had supped sumptuously on boiled eggs and peach preserves, I turned into a large double bed that looked as if it had just been spread for my special accommodation, and with Scout keeping watch at the door I slept like a prince until the sun was an hour high.

For my breakfast I had fried chicken, ash cake, boiled eggs, coffee and honey. After breakfast, I filled my knapsack with fresh provisions, and bidding adieu as I thought forever to these pleasant quarters, I set out again to search for a road to lead me across the brake. Little did I think that five days would pass before I bade a final farewell to these quarters, yet such was the fact.

All that day I searched for a road that would lead me across the interminable cane brake that barred my further progress. Occasionally I would fall into a cattle or deer trail leading into it, but they either gave out entirely after penetrating it a short distance, or else split up into half a dozen blind paths that did not seem to lead anywhere or in any particular direction. Wearied and disheartened by my failure to find a road, I returned to my domicile, feasted again on fried chicken, eggs, honey, etc., and again took possession of my double bed for the night.

The next day this same programme was gone through, and the next, and the next, with the same results, and I almost began to despair of ever finding a way through this apparently endless wilderness of cane, briers and brush. However, it was some consolation to me to know that after the fatigues and disappointments of the day, I had such comfortable quarters to fall back upon at night.

Nevertheless, as I was very anxious to get on as speedily as possible, I left my domicile one morning with the determination that I would follow the brake up to the head of old Caney, providing I could find no road crossing it. I went on up the brake, examining closely every nook and indentation without success, until I had traveled, as I suppose, five or six miles. Here I struck out into the open prairie to avoid a deep lagoon that lay in the way, and ere long I came to a well-beaten road, running almost parallel with the brake. This road had evidently been traveled a day or so previously by a large body of cavalry. I concluded I would follow it a short distance, and was going along leisurely, when I heard the clattering of horses' hoofs behind me, and turning to look, I saw a troop of Mexican lancers advancing rapidly, not more than four or five hundred yards distant. There was not a tree or bush to screen me, nearer than the brake, at least half a mile to my right, and I knew it would be impossible for me to reach it before I was overtaken by the lancers. For a moment I gave myself up for lost, but fortunately on one side of the road there was a patch of rank dead grass and, as there was no time for consideration, I seized

Scout by the neck, dragged him twenty or thirty paces into the grass, threw him down and laid myself by his side, holding him tightly by the muzzle to prevent him from growling or barking at the lancers as they passed.

In a few moments they came up and when opposite the place where Scout and I were hidden, they halted. I could see them plainly through the grass, and could hear them talking, but not with sufficient distinctness to understand what was said. Scout, too, was aware of their proximity, and when they halted he gave a low growl, and tried to get up, but I choked him severely until he lay quiet. The lancers had evidently caught a glimpse of us before we left the road, for after they halted, several dismounted and examined the road for tracks, but luckily at that place the ground was gravelly and hard, and my boots had left no distinct traces on it.

At length, satisfied I suppose that they had seen nothing, or what they had seen was only a couple of wolves or wild hogs, those that had dismounted to examine the road for sign sprang into their saddles, and they all rode on at a gallop. As soon as I saw they were fairly off, I drew a long breath, and I think Scout did so too, for I had choked him until his tongue lolled out, When the lancers had got to a safe distance, I loosened my grasp from his neck and let him up. But he never forgot the lesson I gave him on that occasion, and whenever I wished him to lie down and keep quiet, I had only to place my hand on his neck, when he would crouch down and remain as still as a mouse until I told him to rise.

Thankful for what under the circumstances seemed to me almost a miraculous escape, I took my way back to the timber, resolved that henceforth I would keep a better lookout, and travel as little as possible in daylight through the open prairies. When I reached the woods the sun was about setting and, as it was too far to think of returning to my domicile, I selected a suitable locality and encamped for the night. During the night several large animals which I supposed to be bears came around camp, and the noise they made in the cane kept Scout in such constant state of excitement, that I am sure he got but little sleep.

The next morning, I retraced my way down the brake, and about mid-day reached my quarters, where I found everything as I had left it the day before. After feasting again on fried chicken, sweet potatoes and hot coffee, I took a seat on the porch, with a volume of Don Quixote (which I read for the first time at this house) and, cocking my feet up on the bannisters, I made myself comfortable for the rest of the evening.

Whilst I was thus taking "mine ease in mine inn," it occurred to me that if Mahomet couldn't get to the mountain, perhaps the mountain might come to Mahomet—in other words, if I couldn't get to the Texan army, perhaps it would be just as well to remain where I was until the Texans whipped the Mexicans and reoccupied the country. That they would do so eventually I had not the slightest doubt, although the Mexicans had told us at Goliad (for the purpose of discouraging us and preventing us from making any attempt to escape), that Santa Anna had defeated Gen. Houston's army, and that the whole country was virtually in their possession. But in fact I did not seriously entertain for a moment the idea of remaining any longer where I was, comfortable as were my quarters, than I could possibly help; for I knew very well I would not be satisfied with such an inactive life, when my countrymen were all in the field battling against the merciless foe. So I retired to my sleeping apartment that night with the determination of renewing my search for a road the next morning, and to persevere in it until I succeeded.

During the night I heard the howl of several lobo wolves very near the house, but of course I did not fear them within the walls of my castle. The fact is, I did not fear anything except a visit from marauding parties of Mexicans or Indians, against whom neither the log walls of my castle nor my two formidable looking carving knives would have afforded me much protection. Audubon, who is a recognized authority upon the subject of birds, if not of beasts, told me that the lobo was the largest known species of wolf in the world, and certainly they are much larger than any on the American continent. They resemble the hyena in form as much or more than they do that of the common wolf. Their howl is also very different, and when camping out alone on the prairies, it always seemed to me to be the most mournful, doleful and lonesome sound I ever heard. Several instances have been known since the settlement of Texas of their attacking travelers when benighted on the prairies, and I was once myself with a party of rangers who rescued a wayfarer from their clutches, and who, but for our timely arrival, would undoubtedly have been torn to pieces by them.

Nothing else occurred to disturb me during the night, and the next morning I arose early, and as soon as breakfast was over I shouldered my knapsack and set out, intending to make a thorough search for a road along the edge of the brake below. In the bottom today I noticed that many of the trees were putting forth their leaves, an indication that spring had fairly set in, and a variety of wildflowers were also beginning to make their appearance on the prairie. I came across a specimen of the jointed snake, the first I had ever seen. It was a small snake, not more than fifteen or twenty inches in length, and its skin had a vitrified or glassy appearance. It seemed to be rather sluggish and unwieldy, and when I struck it a slight tap with a small stick, to my great astonishment, it broke into half a dozen pieces, each piece hopping off in a very lively way on its own hook. I have since heard it asserted, that after a time the broken parts of the snake will come together and reunite and then crawl off as if nothing had happened to it; but I shall always be doubtful of the story until satisfactory vouchers of its truth, duly authenticated and sworn to, are produced.

About mid-day I noticed a cloud of dust rising in the prairie way off to my right, caused, as I at first supposed by a large body of troops in motion. I was traveling near the edge of the cane brake, both for greater security and for fear I might pass by without observing it, some road leading across. I therefore quickly concealed myself behind a small thicket, from whence I could see all that was passing on the prairie. Presently I saw issue from the cloud of dust a dense body of horses, which, on their nearer approach, I perceived were uncurbed by bit and riderless. I supposed there were at least six or seven hundred in the drove. I saw they would pass within a short distance of the thicket where I was concealed, and when nearly opposite, I suddenly sprang out in full view of them and gave a loud whoop. They halted at once and with heads erect, stood for an instant looking at me in astonishment, then with the precision of a troop of cavalry, they wheeled about and went back in the direction they had come.

I continued on my way, and when I supposed I had traveled at least six or seven miles from where I had started, to my great joy I came to a plain road, running from the prairie into the brake. I felt confident it would take me through it, but when I followed it a hundred yards or so into the brake, it came to an abrupt termination at a place where a large tree had been cut down and split into boards! There was not a vestige of a road beyond that point—nothing but almost solid walls of tall canes matted together with green briers and vines.

Sadly disappointed and dispirited, I retraced my steps to the prairie, and thence back towards what I began now to regard as my permanent home, where I arrived a little after sunset, so beat out with my day's tramp that I turned into my bed supperless and slept like a log until roused at daylight by the crowing of my chickens and the squealing of my pigs. It may seem strange to some, that one accustomed to walking as I was, and after living upon the fat of the land as I had of late, that I should have been so much fatigued by a little tramp of twelve or fifteen miles—but that was precisely "what was the matter with Hanna." After starving for so long a time, I had indulged too freely in fried chicken; and besides, walking through the woods and prairies is not like traveling on a well-beaten road. In the former your progress is often necessarily slow and laborious on account of having to force your way through rank grass and many creeping vines, that are constantly entangling one's legs, and occasionally tripping one up. Moreover the soles of your shoes soon become as slick as glass by rubbing on dry leaves and grass, so that you are frequently slipping backward instead of going forward.

Being determined to persevere in my attempt to find a road that would enable me to cross the brake, the next morning I shouldered my knapsack, and set out again in the direction I had taken two days previously when I made such a narrow escape from the lancers. Scout evidently seemed to think I was wandering about in a very aimless way, nevertheless he trotted along after me without asking any questions. I traveled up the brake a mile or so beyond the point where I had turned back on the former occasion, examining closely every nook and bend for trails or roads. In this way I discovered one or two that had escaped my observation on my previous trip, but they petered out after going a short distance into the cane.

Finding no road or trail to answer my purpose, and night coming on, I encamped in some timber near the edge of the cane. A little after dark I heard a great many turkeys flying up to roost in the trees around my camp. The wolves howled incessantly, and once the sharp scream of a panther close by roused Scout from his slumber and he dashed off in the direction of the sound, but very soon came running back with his tail between his legs. It was evident he wanted my moral support, but I declined hunting panthers in the night with a carving knife. I felt no fear of them, however, in camp, as I had a blazing fire, which I took especial care to keep well supplied with fuel. I have been told that in India tigers have been known to come up to campfires and seize upon persons sleeping near them. This may be true, but there is no wild beast (with the exception perhaps of the grizzly bear) on the North American continent that will venture so near a blazing fire—at least I have never heard of an instance of the kind during the many years I have lived on the frontiers.

At daylight I was aroused from my slumbers by the clucking and gobbling of turkeys. There must have been several hundred of them upon the trees within fifty yards of where I was lying. One fat old fellow was sitting upon a limb not more than thirty feet from me, strutting and gobbling in the most imprudent way. It seemed to me he knew I was particularly fond of roast turkey and that he was "cutting up his didoes" for no other purpose than to tantalize me with the display of his goodly proportions. Even when I got up and walked towards him, he took no notice of me until I threw a stick at him, when he uttered an exclamation something like "what!?" and soared away to his feeding grounds.

After breakfast I continued my route along the edge of the brake. When I had gone about two miles, I noticed a house on the prairie near a small grove of timber, half a mile or so to my left, and I concluded to go out and examine the premises. The house was a small log cabin, surrounded by an enclosure containing perhaps a dozen or fifteen acres. It was poorly furnished and I saw nothing about the premises except some ducks and chickens.

As I did not know how long it might be before I should have a chance at fried chicken again, I determined to take toll out of the poultry about this house. With the assistance of Scout I soon caught and killed two fat pullets and a duck, which I tied on the outside of my knapsack. I then took a plain road running near the house and nearly parallel with the brake, and when I had gone about a mile I met with an adventure that terminated in the most singular and unaccountable manner.

The road at that point was about a quarter of a mile from the brake. How it happened I did not see them sooner, I cannot imagine, unless I had fallen into what the negroes call a "fit of the mazes," but at any rate I suddenly found myself nearly opposite to two Mexican soldiers who were seated on the grass about forty paces to the left of the road. One of them was armed with a musket and the other with a lance, similar to those I had seen used by Mexican cavalry. Near them a horse, saddled, was grazing, and one of the soldiers held the end of his lariat in his hand. I have since thought the horse must have been lying down until I came near them, as otherwise I think I would have seen him sooner.

As I have stated, it was a quarter of a mile at least to the nearest part of the brake, and the idea flashed across my mind that after all my narrow escapes I was certainly caught at last. Retreat to the brake I knew was impossible, as they could easily overtake me on the horse, and for a moment I stood irresolute not knowing what course to pursue. But the very hopelessness of the case produced a feeling of recklessness as to consequences, and I leisurely continued my way along the road, at the same time trying to look as unconcerned as possible and as if I didn't know (and didn't care) that a Mexican soldier was within five miles of me. All the while however, I was watching them closely. As I passed them, they made no movement except to turn their heads and gaze at me apparently in the utmost astonishment, which considering the figure I cut, just at that time, is not to be wondered at. There is not the slightest doubt that I presented a very singular and anomalous appearance. I was tanned by long exposure to sun and weather until I was nearly as dark as an Indian; my cap resembled a Turkish turban, the leather front having been long since carried away in some of its frequent encounters with green briers and other thorny shrubs; my hunting shirt was ragged and blackened with smoke, and my pantaloons, or what remained of them, were buttonless, and held up by a broad leather belt, from which a tin cup hung dangling on one side and two long carving knives on the other, and to complete this unique costume, my shoulders were surmounted by a portly knapsack, to which were tied the two pullets and the duck I had just killed.

This *tout ensemble* of course accounts reasonably enough for the astonishment with which the soldiers gazed upon me as I passed, but still it does not satisfactorily explain their subsequent movements, especially as they could plainly see that with the exception of my two carving knives, I had no arms. However, they did not move until I had gone forty or fifty yards beyond them, when both suddenly rose to their feet and hastily mounted their horse, one behind the other. I, of course, supposed they intended to pursue me, but to my great wonder and astonishment as well as relief, they went off in the opposite direction, across the prairie, as fast as they could urge their horse on with whip and spur. The one mounted behind had a short heavy whip called a "quirt," and as far as I could see them distinctly, his quirt was incessantly and vigorously applied to the flanks of their steed, and every now and then I could see them looking back as if they expected me to pursue them.

What they took me for I am at loss to imagine, but if they had taken me for Old Nick himself I would not have quarreled with them on that score, in consideration of the expeditious manner in which they had left the field—not staying even to say *adios*.

For fear, however, I might not prove to be such a terrible object to other straggling parties of Mexicans whom I might possibly meet with on this road, I left it, and did not halt until I came to the brake. There I stopped to rest a while and hold a council of war with Scout as to what was to be done next. Scout, although he expressed no opinion on the subject, I know was strongly in favor of going back to the "flesh pots of Egypt," and finally we agreed to return to our old quarters. I had noticed an old axe there in one of the outhouses on the place, and I determined to set to work regularly and cut my way with it through the brake, if it took me a month to do it. It seemed very strange to me at the time, that the settlers on old Caney did not cut roads through it when they retreated before the Mexican army. But subsequently, when I mentioned the matter to one who lived on Caney when the settlers abandoned their homes there, he told me that all living on the south side had cut roads from their houses across the brake, but that in every instance they had some circuitous way to reach them, and that no sign of a road was visible on the edge of the brake. This statement was confirmed to some extent by the fact that no one, unless closely searching for it as I was, would have suspected the existence of a road where I found one.

In pursuance of the course I had determined to follow, after resting a while, Scout and I started back to our old quarters, and about an hour before sunset I crossed the strip of cane and the bridge of logs over the bayou and entered the little prairie in which my domicil was situated. As I was proceeding leisurely towards the house, it occurred to me that it might be well to examine again the north side of the prairie bordering the main brake which heretofore I had only partially done. With this intention I left the path I was following, and when I had gone a few hundred yards I came to a trail leading towards the brake along which the marks of wagon wheels were dimly visible. This I followed until it led me into an indentation in the brake, which was so narrow and so well concealed by bushes and cane as to be barely perceptible at the distance of a few paces. Still following the traces of wagon wheels, I came on the farther side of this nook to a newly cut road wide enough for the passage of a wagon and team.

I was satisfied that at last I had found what I had been so long in search of, but in order to assure myself of the fact, I followed the road for nearly half a mile into the brake, and as it still ran on in the same direction I was convinced it would take me through. By this time the sun had set, and I concluded to return once more to my old quarters, and make an early start in the morning.

As I walked along my attention was suddenly drawn to two large black objects in the road a short distance ahead of me. I stopped a moment to ascertain what they were, and as I did so, Scout gave a low growl and retreated behind me. By the dim light that struggled through the overlapping canes, I at length discovered that these black objects were two huge bears, standing perfectly still in the road, and apparently waiting for us to come up. For an instant I thought of retreating, but on reflection, as I had never heard of anyone being attacked by black bears unless wounded, I screwed up my courage (nearly breaking the screw-driver in the attempt) and resolved to pass them if I could. There was no chance to go around them, for the cane was so thick on both sides of the road, I might almost as well have tried to penetrate a solid wall. So I drew my longest carving knife, and boldly (apparently) advanced towards them. They stood perfectly still until I was within eight or ten feet of them, when they commenced growling, and looked so large and ferocious, and so bent on disputing my right-of-way, I felt more than half inclined (as Scout had done already) to tuck my tail and beat a hasty retreat.

But I knew it was too late to turn back, and that any show of timidity would embolden them to attack if they had not intended doing so. I therefore continued to advance, and my apparent boldness seemed to daunt them a little (if they had only known how badly I was scared I am sure they would have seized me) and when almost near enough to have touched them, one of them sullenly drew off to one side of the road and one to the other, and Scout and I passed between them. As we went between them, they showed their white teeth and growled so fiercely that every instant I expected they would rush upon us, but they did not, nor did they attempt to follow us. All the while Scout kept close at my heels with his tail between his legs—the first and last time I ever saw him completely cowed.

It is asserted that the black bear never attacks a man, unless wounded or brought to bay, and I do not say positively that these two had any intention of making their supper on us, but to say the least of it, their bearing towards us was exceedingly suspicious; and besides, I thought they might just as well kill a fellow at once as to scare him to death. At any rate Scout and I congratulated ourselves (at least I know I did), when we were once more safe within the four walls of our house.

I rose early next morning to prepare as much provision for the road as I could conveniently carry. I cooked the duck and one of the pullets I had killed the day before (Scout and I had demolished the other for breakfast), and ground a gallon or so of meal on a steel mill. Besides these, my supplies consisted of five or six pounds of bacon, several pounds of sugar, two pounds of coffee parched and ground, some salt and pepper, and two bottles of honey. This, I thought, with care would last us eight or ten days, even if we found nothing, on the road. I also had a tin cup for making coffee, and of course my two carving knives which I had sharpened on a whetstone were as keen as razors. For these I had made scabbards out of a piece of leather and sewed them to my belt. When ready to start I scribbled with a bit of charcoal the following "due bill" upon the wall of my sleeping apartment:

> *John Duval, an American captured by the Mexicans but escaped from them at Goliad, is indebted to the proprietor of this house for one week's board and lodging and some extras, and will pay the same on demand.*

The extras referred to consisted of articles of clothing, pipe and tobacco, etc. This note has never been presented for payment, and I suppose it is barred now by the statute of limitation—nevertheless, I would cheerfully pay the principal now, but not the interest, for that would put the amount far above my present assets, and I should be compelled to take the benefit of the Bankrupt Act. Having thus settled my board bill on such easy terms, I shouldered my knapsack, stuck my carving knives into my belt, and followed by Scout, I took my way towards the road I had found the evening before.

Just as I was entering the brake, I turned to take a last look at the house that had been a haven of rest to me after my wanderings in the wilderness, and I experienced a feeling of regret when I thought that in all probability I should never see it again. There I had truly been "the monarch of all I surveyed." I could loll upon the sofas—tumble up the beds—wipe the mud from my boots on the rugs and carpets—smoke tobacco (by no means of the best quality) in the drawing room—select my own "menu" from the well stored pantry and the poultry in the yard—and there was none to say me nay. Even now I look back with pleasant recollections to my sojourn in those comfortable quarters, for it was the only time I ever had complete and undisputed control of such an establishment. Peace to its ashes if, as is highly probable, it was subsequently burned by the Mexicans.

As I passed the place where I had encountered my doubtful friends the two bears the evening before, I noticed many of their tracks in the mud on the side of the road. They were very much like the tracks made by a barefooted man (no pun intended for I detest puns), except that the heel part was as long as the toe. After traveling I suppose between three and four miles, crossing on the way a sluggish bayou, over which I "cooned it" on a fallen tree, to my great satisfaction I saw light ahead, and in a short time came to the open prairie.

At that time, as I have before stated, nearly the whole of the bottoms on old Caney was covered by an unbroken canebrake some sixty or seventy miles long and from three to five in width. This I had from others who were settlers in that portion of the country at an early day, and the statement is probably correct. The soil of this brake is exceedingly fertile, and the time will come no doubt, when it will be converted into one continuous sugar and cotton plantation. At the point where I saw it, it was a dense mass of cane, briers and vines, with here and there a scattering tree growing in their midst. Bears, panthers, wild hogs and other varmints were very numerous in it and along its borders.

About half a mile below the place where I came out into the open prairie, I saw a house near the bottom, and as I had made it a rule to search every one I passed for guns and ammunition, I started with that intention towards the one in question. I kept well under the shelter of some timber bordering the brake, to screen myself from the view of anyone who might be about the premises. In this timber I struck a plain trail leading towards the house, which I took. I followed it perhaps a hundred yards or so, when as I turned a short bend in the path, I caught sight of a Mexican soldier, with his gun on his shoulder, walking rapidly towards me. Luckily a dense growth of bushes bordered the path at the point where I then was, and although I had but little hope the Mexican had not seen me, I instantly sprang into the bushes and laid down among them. Scout, who evidently had not forgotten the choking I gave him on a previous occasion, quickly followed me, and took his station by my side. It seems, however, the Mexican did not notice us, for he came on and passed within six feet of us without halting. I could almost have touched him with my longest carving knife, and if he had been a little weakly chap I think I would have been tempted to spring suddenly upon him as he passed and give him a tussle for his gun. But he was a big strapping fellow, and I knew I would have no chance of coming off the winner in a hand to hand encounter with him, even if I had not been hampered with a heavy knapsack, and other impediments. I concluded therefore that "discretion was the better part of valor," and did not move until he was hidden from my view by a turn of the path.

As it was evident he came from the house I had seen, and as I thought it highly probable there were more of the same sort there, I gave up the idea of searching it for guns, for fear I might find more of them there than was desirable; so I gave it a wide berth, and striking off through the woods to the right I came out again to the prairie two or three miles below.

The day was cloudy and dark, and I couldn't see the timber on the opposite side; consequently I could form no idea of its extent. Besides (having made a late start on account of being delayed in preparing provisions for the road), the sun was by this time getting pretty low, and I thought it best to encamp for the night and start anew in the morning.

In a little open space just within the brake, separated from the prairie by a very narrow strip of cane I pitched my camp; in other words, I pulled off my knapsack, and stretched myself upon a bed of dry grass which I had cut with a knife. It was too early to cook supper, and as I had no dread of wild beasts till dark, I did not start a fire, and very fortunate it was for me I had not done so. I was just falling into a doze, when Scout gave a low growl and at the same moment I heard the tramping of horses' hoofs. I looked through an opening in the strip of cane between me and the prairie and saw five or six Indians who were driving a number of horses, coming along the edge of the brake. Just as they were opposite to the spot where Scout and I were lying, two of the horses broke away from the caballada, ran through the strip of cane and nearly over us. One of the Indians started after them, and was crossing the strip of cane, when the two runaways seeing Scout and I lying upon the ground, suddenly wheeled and ran back to the prairie, and the Indian turned also and followed them. If he had come six feet further, he must inevitably have seen us. As it was, he did not discover us and the Indians and their drove of horses soon passed out of sight.

These two close calls both occurring the same day convinced me that I had but little chance to make my way safely through a country swarming with roving bands of Mexicans and Indians; and yet, although I passed their recent encampments at several places, I never saw an Indian afterwards, nor a Mexican, except some squads of cavalry a long way off on the prairie.

During the night I heard bears crashing through the cane, and splashing in the water of the pool near which I was encamped. The number of bears at that day on old Caney was so great I cannot imagine how the settlers there managed to raise hogs unless they kept them constantly penned up. The next morning I saw many of their tracks on the edge of the pool, where they had been digging up some kind of plant with a bulbous root.

The next morning as soon as I had eaten breakfast and Scout had cleaned up the fragments, I set off towards a long line of timber that was just barely visible on the farther side of the prairie. Not a great while after I had started I noticed a long way off to the west, a column of smoke rising up, which I supposed indicated an encampment of Mexicans or Indians in that quarter. When I had traveled perhaps three or four miles, I observed that this smoke was increasing rapidly in volume and extent, and that it appeared to be approaching the direction I was going. Then, for the first time it occurred to me that the prairie was on fire, and I began to be seriously apprehensive that the fire might overtake me before I could reach the timber. The grass of the last season's growth was from fifteen to eighteen inches in height, and as dry as tinder, and it seemed very probable, with the stiff breeze blowing at the time, that the fire would overtake me before I could gain the opposite side of the prairie, still five or six miles distant.

I hurried on as fast as I could, but before I had gone two miles further, I was convinced that escape by flight was impossible. I had heard old frontiersmen say, that the only thing to be done in a case of this kind, was to fight fire with fire. I took my flint and steel from my pocket, ignited some tinder which I wrapped in a wisp of dry grass, and swinging it quickly backwards and forwards in my hand, it was soon in a blaze. With this I set fire to the grass ahead of me, and in a few moments I had the satisfaction of seeing my counter-fire sweeping the grass that grew in the direction I was going.

By this time the wall of fire extending in a long line across the prairie behind me was swiftly moving towards me. Already I could see bright tongues of flame flashing out at intervals through the dense column of smoke, and a dull continuous roar, like the distant beating of surf on a rockbound shore, was distinctly audible. Hundreds of deer, antelope and other animals came scampering by me in the wildest terror, and numerous vultures and hawks were seen hovering over the smoke, and occasionally pouncing down upon rabbits and other small animals, roused from their lair by the advancing flames. The nearer it came the faster it seemed to come, and I could see blazing tufts of grass borne along by the wind setting fire to the prairie sometimes fifty or a hundred yards ahead of the main fire. But by the time it had reached the place where I had set my counter-fire going, the grass for several hundred yards was burnt off, and of course the fire was arrested there for want of fuel. I had nothing to do but follow the track of the fire I had started, which cleared the way before me as I went, and rendered walking much less fatiguing than it otherwise would have been, verifying the truth of the old saying "that it is an ill wind that blows no good."

About two hours after I had set my counter-fire going, I came to the outskirts of the timber for which I had been steering, and through which I continued my course until I was stopped by a deep bayou. On the bank of this bayou, in a little open space not twenty feet square, I pitched my camp, and from the fallen trees around I collected fuel enough to keep my fire going all night. There I soon prepared a meal from the provisions I had in my knapsack, to which I and Scout did ample justice as we had not tasted food since early in the morning. As it was still several hours till night, I employed myself in repairing my dilapidated wardrobe with a needle and some thread I had found in my house on old Caney.

Nothing occurred to disturb my slumbers during the night. The next morning after breakfast I shouldered my knapsack and started again. The bayou on which I had camped, though the current was very strong, looked so narrow I thought I could easily swim it without taking off my knapsack; so I plunged in at once, but unfortunately when I had about reached the middle of the stream, one of the straps that held it in position gave way, and in an instant the rapid current twisted it around my neck, and I went down with it like a stone to the bottom. I exerted myself to the utmost to free myself from it but without success, until I thought of my carving knives. With great difficulty I drew one of them from the scabbard (it seemed to me that everything about me was tangled up) and cut the strap that fastened the knapsack around my neck. The moment I was freed from it, I rose to the surface, puffing and blowing like a porpoise, and half strangled with the water I had swallowed much against my will, for I was not in the least thirsty. Scout having no knapsack to encumber him, had already reached the opposite shore, and was running up and down the bank, whining most dolorously, and showing plainly his anxious concern for my safety. I quickly gained the shore myself after coming to the surface, but alas! I was compelled to leave my precious knapsack which contained our whole supply of provisions, at the bottom of the bayou. However, I was very glad to get out of the scrape as well as I had done.

The first thing I did after reaching the shore was to examine the condition of my tinder, and I was glad to find that but little water had penetrated the greased cloth in which it was wrapped. I took it out and spread it in the sunshine, so that what little moisture it had imbibed might evaporate. If I had lost my tinder as well as my provisions, I would have been in a truly pitiable condition.

When I had partially dried my clothes, I set out again in my usual direction, which led me for some distance through a thick growth of underbrush, from which I finally emerged into open post oak woods. I went on through these until nearly sunset, when the howling of wolves warned me that it was time to select a suitable place to encamp. I chose a spot in a thick grove on the margin of a pond. There I started a fire, and as I had to go to bed supperless, I determined that at any rate my bed should be a good one. With one of my carving knives I cut a quantity of long dry grass which I spread before the fire, on which I and Scout, after the mishaps and fatigues of the day, slept soundly till morning.

As soon as it was daylight, as I had no breakfast to cook and eat, I was on my way again, and in a little while I came to a prairie, on the farther side of which I saw a forest and a large lake near it. Towards this lake and forest I steered my course, but after traveling some distance, I was astonished to find that apparently they were as far off as when I first saw them. Whilst I was wondering at this, I noticed that the lake and forest were each moment growing more indistinct, and at length they vanished altogether, and in their place nothing was visible but the level expanse of the open prairie. I knew then that the appearance of this lake and woods was an optical illusion termed a "mirage," produced by some peculiar state of the atmosphere. I have frequently seen them since on the plains in the west, and on several occasions have been cruelly tantalized when suffering from thirst, by the sight of lakes that disappeared before I could reach them.

After traveling a while longer, I saw some distance ahead of me a grove, and still further on a forest was dimly visible. At first I thought it probable that these also were only the ghosts of a grove and forest, and that they too would disappear and give me the slip, but they proved to be the genuine articles. To this forest I steered my course, guided by the intervening grove. I saw several squads of Mexican cavalry on the way, but they did not come near me, and I avoided observation simply by lying down on the ground, until they had passed by. But what astonished me much was, that these squads were all traveling in a disorderly manner towards the west. It soon occurred to me, however, that the Mexican army must have met somewhere with a signal defeat, and that those I saw were straggling detachments from their routed forces. I have no doubt this supposition was correct, for the Battle of San Jacinto, in which Santa Anna was taken prisoner, was fought and won by the Texans under Gen. Houston a few days previously.

About noon, I came to the grove that had served me as a landmark to guide me on my course, and feeling somewhat fatigued, I laid down just outside of it to rest a while. I had been there but a few moments when I had practical evidence that the vast distance at which the buzzard is said to see a carcass on the ground, had not been exaggerated. When I laid down not a buzzard was in sight, although I had an unbroken view for miles in every direction, but in less than five minutes, half a dozen of them were wheeling and circling above my head, and coming lower and lower, evidently for the purpose of ascertaining if Scout and I had been killed long enough to suit their fastidious taste. "My friends," said I, "on this occasion you are a little too 'previous'—you have come very near several times having the satisfaction of picking my bones, but to prove to you that I am not as yet a fit subject for a postmortem feast, I'll move on." The first movement I made satisfied them on that point, and they departed as quickly as they had come.

Continuing my course, about sunset I came to a deep and rapid stream, which I know now was the San Bernard, and I encamped for the night on the bank. By this time I was suffering much from hunger, but there was nothing in camp to eat, and I and Scout were compelled to satisfy the cravings of our appetites as well as we could, by going to sleep. The poet calls sleep "tired nature's sweet restorer," and under ordinary circumstances, no doubt there is some truth as well as poetry in the saying, but when a fellow has had nothing to eat for several days, and his bed is the naked ground, sleep as a restorer isn't a marked success—at least I was just as tired and hungry when I woke up the next morning as I was when I laid down. However, I was in hopes that I might find a settlement on the other side of the river where something to eat could be had, and without any preparation except simply tying my cap on my head securely to keep my precious tinder from getting wet, I plunged into the turbid stream closely followed by Scout. The water was very cold, but I soon crossed over and ascended the bank that rose up almost perpendicularly thirty or forty feet on that side of the stream.

When I got to the top of the bluff, I discovered a house a few hundred yards above me, to which I turned my course. As it was all open prairie on that side of the river except a few scattering groves, I had a good chance to reconnoitre the premises before approaching, and seeing nothing to indicate that the house was occupied, I went up. It proved to be a single log cabin, in rather a dilapidated condition, and had been ransacked by some plundering party of Mexicans who had taken or destroyed any provisions that might have been there, except a handful of corn I found in a barrel. As I was thoroughly chilled after swimming the river, I concluded I would build a fire in the chimney for the double purpose of drying my clothes and parching the corn I had found.

There was but one door and one window to the cabin, both on the same side, and while I was busily engaged in parching corn, my attention was drawn to a grating sound in the direction of the window, and turning to look, I saw the muzzle of a gun protruding through it. But Scout had noticed it, too, and giving a savage growl, he sprang at one bound through the window, and at the same instant almost I heard someone rip out an oath in good, King's English, and exclaiming "Come take your dog off!" in such choking accents as convinced me there was urgent need of haste. I ran out immediately, and with some difficulty forced Scout to let go the grip he had taken upon a thick woolen comforter, which fortunately for him, my visitor had wrapped around his neck.

After he had somewhat recovered from the surprise and alarm into which the unexpected onset of Scout had thrown him, he asked me where I was from, and how I came to be out there all alone among the Mexicans and Indians. When I had satisfied him on this point, he told me that he and a Capt. D were out on a spying expedition, and seeing the smoke coming out of the cabin chimney where I was carrying on my culinary operations, they had come to the conclusion that a party of Mexicans had halted there. After a consultation as to the best mode of proceeding, it was determined that Capt. D should remain with the horses under cover of a grove a few hundred yards from the cabin, whilst his companion, Mr. H, should cautiously approach it on foot, and ascertain the strength of the party within. If too strong for them to contend with, he was to fire upon them through the door or window and then make his retreat as fast as possible to the grove where he had left Capt. D and the horses. But in arranging this programme, they did not consult Scout, who revenged himself in the manner I have stated. After giving me this information and telling me that the Texans had whipped the Mexicans at San Jacinto, etc., Mr. H gave a whoop (the preconcerted signal for Capt. D to come on), and in a few moments he rode up, leading H's horse and another one, which to my great satisfaction I found was well packed with provisions.

As I have stated, I already had a fire underway, and in a little while a pot of coffee was simmering on it, and a haversack of eatables, biscuits, potatoes, cold ham, etc., was spread upon the floor. Those biscuits! I shall never forget them! None of your little thin flimsy affairs, such as are usually seen on fashionable tables, but good solid fat fellows, each as big as a saucer, and with dark colored spots in the center, where the shortening had settled in the process of baking.

When the coffee was ready I was invited to dig in, which I did promptly and without any pressing, after casting a contemptuous look towards the little pile of parched corn on the hearth, which I had previously prepared for my breakfast.

As well as I remember, I think I was dealing with my fifth biscuit, and was looking longingly toward the sixth, when Capt. D mildly suggested that in his opinion I had better knock off for a while for fear of consequences. To this I made no reply except to seize the sixth biscuit, and while I was disposing of that, Capt. D expeditiously cleared the board, and deposited the remainder of the provisions in the haversack. We then mounted the horses (the pack animal having been turned over to me) and in a day or so we reached the Brazos, where a portion of the Texan army was encamped.

I have nothing further to add, except that when I left for the States a month or so subsequently, finding it impossible, owing to the crowded condition of the schooner in which I sailed to take Scout with me, I gave him to my friend H, who promised me he should be well taken care of. Many years afterwards I met with H. at Austin, and he told me that Scout lived to a good old age, and died the respected progenitor of a breed of dogs that were highly prized for their valuable qualities…

Portilla at Goliad to Urrea at Victoria, the day of the massacre:

My Dear General:

I feel much distressed at what has occurred here; a scene enacted in cold blood having passed before my eyes which has filled me with horror. All I can say is, that my duty as a soldier, and what I owe to my country must be my guaranty. My dear General, by you I was sent here; you thought proper to do so, and I remain here in entire conformity to your wishes. I came, as you know, voluntarily, with these poor Indians to cooperate, to the best of my humble means, for my country's good. No man is able to do more than is within the scope of his abilities; and both they and myself have doubtless been placed here as competent to the purposes you had in view. I repeat, that I am perfectly willing to do anything, save and excepting the work of a public executioner by receiving orders to put more persons to death. And yet, being but a subordinate officer, it is my duty to do what is commanded me, even though repugnant to my feelings.

I am, General, your devoted and sincere friend,

J. N. de la Portilla

www.ingramcontent.com/pod-product-compliance
Lightning Source LLC
Chambersburg PA
CBHW020338100426
42812CB00029B/3168/J

9781941324028